W9-DBH-148

BD
111
.S575
1995

DISCARD

Time, change and freedom

This is no ordinary introduction to metaphysics. Written for the most part in an engaging dialogue style, it covers metaphysical topics from a student's perspective and introduces key concepts through a process of explanation, reformulation and critique.

Focusing on the philosophy of time, the dialogues cover such topics as the beginning and end of time, the nature of change and personal identity, and the relation of human freedom to theories of fatalism, divine foreknowledge and determinism. Each dialogue closes with a glossary of key terms and suggestions for further reading.

Time, Change and Freedom concludes with a discussion of the metaphysical implications of Einstein's theory of relativity and a review of contemporary theories of time and the universe.

Written throughout in an accessible, non-technical style, *Time, Change and Freedom* is an ideal introduction to the key themes of contemporary metaphysics. It will be invaluable for all students on introductory philosophy courses as well as for those interested in metaphysics and the philosophy of time.

Quentin Smith is Professor of Philosophy at Western Michigan University. **L. Nathan Oaklander** is Professor of Philosophy at the University of Michigan–Flint.

Time, change and freedom

An introduction to metaphysics

Quentin Smith
and L. Nathan Oaklander

London and New York

To the memory of Jacinto R. Galang, Jacinto A. Galang and Romy Galang; and to Janet and Howard Smith, necessary conditions of the possibility of Part I and the Appendix.

First published 1995
by Routledge
11 New Fetter Lane, London EC4P 4EE

Simultaneously published in the USA and Canada
by Routledge
29 West 35th Street, New York, NY 10001

© 1995 Quentin Smith and L. Nathan Oaklander

Typeset in Palatino by Florencetype Ltd, Stoodleigh, nr Tiverton, Devon

Printed and bound in Great Britain by
T. J. Press (Padstow) Ltd., Padstow, Cornwall

All rights reserved. No part of this book may be reprinted or reproduced or utilized in any form or by any electronic, mechanical, or other means, now known or hereafter invented, including photocopying and recording, or in any information storage or retrieval system, without permission in writing from the publishers.

British Library Cataloguing in Publication Data
A catalogue record for this book is available from the British Library.

Library of Congress Cataloging in Publication Data
Smith, Quentin, 1952–
Time, change and freedom: an introduction to metaphysics/Quentin Smith and L. Nathan Oaklander.
p. cm.
Includes bibliographical references and index.
1. Metaphysics. 2. Time. 3. Change. I. Oaklander, L. Nathan, 1945–. II. Title.
BD111.S575 1994
110–dc20 94–34474
CIP

ISBN 0–415–10248–0 (hbk)
ISBN 0–415–10249–9 (pbk)

Contents

Introduction

Quentin Smith

The word "metaphysics" has become a part of popular culture and almost everybody thinks they know what "metaphysics" means. This is unfortunate for philosophers, for the popular meaning of "metaphysics" is very different from the philosophical meaning. Popular metaphysics deals with such topics as "out-of-the-body experiences," levitation, astral projections, telepathy, clairvoyance, reincarnation, spirit worlds, astrology, crystal healing, communion with the dead and other such topics. Popular metaphysics consists of notions that for the most part are inconsistent with science or reason. Private and unverifiable experiences, fanciful speculations, hallucinations, ignorance of science and the misuse of logical principles are typical of the ingredients found in popular metaphysics. Given the difficulty and the enormous time and effort it requires to think in a logically systematic way and to understand current science, it is not surprising that more people are attracted to popular metaphysics than to philosophical metaphysics.

Philosophical metaphysics, the subject of this book, is at the far end of the spectrum from popular metaphysics. Philosophical metaphysics is both consistent with, and in part based upon, current scientific theory, and it uses logical argumentation to arrive at its results. For example, if current science informs us that the universe began to exist 15 billion years ago with an explosion called "the big bang," then metaphysics will take this theory into account in formulating theories about the beginning of time and the universe. Moreover, philosophical metaphysics takes logical consistency as a necessary condition of truth. In popular metaphysics, one can say "I don't care if there is a logical disproof of my theory; I still believe my theory because I feel in my heart that it is true." But one cannot get away with this in philosophical metaphysics; if one's theory has been shown to be logically self-contradictory, then one abandons the theory.

Philosophical metaphysics aims to answer two sorts of questions: (1) What is the basic nature of reality and what are the basic kinds of items that make up reality? (2) Why does the universe exist?

The question about the basic nature of reality has usually been called "ontology," after the Greek word *ontos* (beings). Ontology is the study of beings, the study of What Is. The question about why the universe exists has for centuries been regulated to a second area of metaphysics, "philosophical theology," after the Greek work *theos* for divinity. In the present book on *Time, Change and Freedom* we shall deal with ontology. In another volume, *Theism, Atheism and Big Bang Cosmology*, by W. L. Craig and Q. Smith,[1] philosophical theology is addressed, the study of why the universe exists and whether or not there are reasons to believe there is a divinity. In this second volume, I have shown that there can be an answer to the metaphysical question, "Why does the universe exist?" that does not appeal to any divinity, but to certain laws of nature, such as the laws of nature that Stephen Hawking has discussed in his book, *A Brief History of Time*.[2] Since the question of why the universe exists has been for centuries associated with the question of whether God exists, it has come to seem natural to associate the words "philosophical theology" with this branch of metaphysics. A more neutral title of this branch might be "explanative metaphysics," the branch of metaphysics that attempts to determine if there is an explanation of why the universe exists.

The present volume is on ontology. What are the general features of reality, what sorts of beings make up reality and how are the various sorts of beings related to each other? From the very beginning, time has played a central role in ontological studies. Perhaps the earliest influential metaphysical theory was Plato's. Plato divided reality into two sorts, depending on how it stood in relation to time. For Plato, true being or full being belongs to the everlasting, the permanent, whereas imperfect being, the impermanent, belongs to the realm of what comes to be and passes away in time. This reliance on time to divide basic categories of being became even more prominent in medieval metaphysics, where concepts related to time, specifically eternity, were understood as the paradigm of being itself. "To be" in the full and perfect sense is to be eternal, and anything else "is" at all only to the extent that it imitates the eternal mode of being. With modern thinking, we find less emphasis on eternity, but more emphasis on time as the central feature of reality. We find Kant saying that time is the fundamental way in which the mind understands reality, and in twentieth-century existential theory we find Heidegger saying that time is the meaning of Being itself.

Time plays an equally fundamental role in twentieth-century analytic metaphysics, which is the metaphysical tradition to which the present book belongs. We shall take time as the key to our entry into metaphysics and as the unifying theme of our discussions of the various sorts of beings. The understanding of the nature of substances, events, persons, changes, eternity, divine foreknowledge, fatalism, the universe, all require that we understand how these kinds of items are characterized in terms of their temporal characteristics or their relation to time. If we want to understand the universe, we want to know if time began to exist or whether the past is infinite. If the universe began to exist, can there be empty time that elapsed before the universe came into existence? Questions about everyday objects also involve temporal notions. The understanding of things, events, changes, personal identity, free will and the like also requires an understanding of their relation to time. For example, a substance (such as a table) is a thing that endures through successive times, whereas an event is often understood as a substance possessing a property at a certain time. A change is a substance having one property at one time and losing that property at a later time. Metaphysical issues about freedom also involve time; for example, the question about whether we have free will and about whether we are "fated" to live our future lives in a certain way depends on how present time is related to future time. Indeed, the very questions, "What is reality?" or "What is being?" cannot be answered without bringing in the notion of time. For example, does reality consist only of the fleeting present, what is occurring now? Or is reality extended into the future and the past, such that the future and the past are equally as real as any time we choose to call "the present"? Furthermore, does reality divide into two realms, the eternal and the temporal, or does reality consist only of time and its occupants?

These and other metaphysical questions are addressed in the three parts of this book and the Appendix.

Part I has the title "The finite and the infinite" and deals with issues pertaining to the finitude or infinitude of time. Dialogue 1 discusses the beginning of time; Dialogue 2, infinite past and future time, and Dialogue 3, the question of whether time consists merely of the events in the universe or is an independent "substantial" reality that would continue to "flow on" even if there were no events. In Dialogue 4, we shall consider several definitions of divine eternity and discuss whether or not it is possible for there to be an eternal being.

In Part II, "Time and identity," we shall turn to issues of change, temporal passage and personal identity. Things persist and retain their identity through time and change. But what is identity through time? What is change? These questions are addressed in Dialogue 5. Dialogue 6 on the passage of time addresses the issue of whether The Present is the sole reality or whether all times (be they 500BC, 1994 or 1999) are equally real. Dialogues 7 and 8 are about the nature of personal identity and how personal identity is related to time.

In Part III, "The nature of freedom," three issues about freedom are discussed. In Dialogue 9, there is a discussion of whether or not we are fated to live our future lives in a certain way. Dialogue 10 considers if it is possible for us to make free choices if God exists and foreknows every choice we will make. In Dialogue 11, there is a discussion of whether or not all our decisions are caused by past events and, if so, whether that is compatible with our decisions being "free" in some sense.

The Appendix, "Physical time and the universe," addresses topics that are closely related to current scientific theories, especially Einstein's Theory of Relativity and current cosmological theories. Physics, and especially physical cosmology, has developed extensive theories about time and the universe that are confirmed by the observational evidence. The task of the metaphysician is to interpret the significance of the physicist's equations for an understanding of time and the universe. Many of the most unusual, or possible, features of the universe are originally based on physical theories rather than purely philosophical theories. We shall analyze the meaning of the scientific theories of time-travel into the past, of splitting universes, theories that our universe occupies a time series that branches off from a background "trunk-time," that time could be closed like a circle, that what is real and present is relative to one's reference frame, and related topics.

What will the reader learn from this book on metaphysics? The reader will not learn something in the same sense that she might in reading a textbook on chemistry or biology. There is no established body of knowledge in metaphysics. On virtually every subject, there is widespread disagreement among metaphysicians. One reason for this difference between science and metaphysics is that scientific theories lead to predictions of observations that can be used to settle disputes. For example, if one theory predicts that the earth revolves around the sun, and another theory predicts that the sun revolves around the earth, then there is a way to resolve the dispute, for example, by observing the sun and earth from the vantage point of a rocket in outer space. However, the subjects that are studied in metaphysics do not lead to predictions of observations and consequently, disputants in this field must rely on logical argument from premises and try to demonstrate logical fallacies in the argument of their opponent. The opponent typically responds by arguing that some of the premises are false, or by claiming that the conclusion does not follow from the premises, or by revising his own theory to render it immune to the argument. This process of argument and counterargument tends to go on indefinitely; consequently, progress in metaphysics is measured not by definitive results but by the increasing sophistication of the theories that defenders of opposing positions develop.

But this is not to say that there are no right answers in metaphysics. There *are* right answers, but the issues are so complex and difficult to resolve that it is extremely hard for us humans – fallible products of evolution that we are – to arrive at definitive and universally accepted answers to the questions. Perhaps the problem is that the human species is intelligent enough to ask metaphysical questions and to develop arguments for certain metaphysical positions, but not intelligent enough to provide definitive answers to these questions. Some may conclude from this that we should concentrate on questions that are easier to answer, such as scientific questions, but this conclusion is unworkable. It is unworkable because *we cannot help but adopt and live by various metaphysical beliefs*. For example, we must live as if there is an eternal God or as if all that exists is matter and organisms that exist in time. And we must live as if we have free will or as if we are "fated" to do everything we do. Metaphysics deals with the rock-bottom issues that no one can escape, unless they live their lives in a coma. The only choice is to adopt a sophisticated and well thought out metaphysical theory or to accept glib and simple-minded answers to the questions. The point of this book is to stimulate the reader to develop a well thought out metaphysical theory on the various metaphysical topics discussed in this book. What will be *learned* from this book is not a body of truths but rather a set of arguments and counterarguments for various metaphysical positions. For example, the reader will learn some of the main arguments for the thesis that we have free will and some of the main arguments that we do not have free will. We hope that the reader will assess these arguments and counterarguments on her own and try to make up her own mind on each of the issues.

The disputational nature of metaphysics explains why we have adopted a dialogue form for Parts I to III of this book. A dialogue presents metaphysics in its true nature, as a sustained debate between opposing philosophers on each of the various metaphysical topics. To present arguments for only one side of the issue, as do many metaphysical books, creates an impression of bias and disguises the truly controversial nature of the subjects.

However, we do not use the dialogue form in the Appendix. The Appendix explains, interprets and draws conclusions from Einstein's Theory of Relativity and current physical cosmology. As we draw closer to the sciences, there is less room for debate and disagreement. There is widespread agreement about the fundamentals of Einstein's theory and hence a dialogue or debate form would be inappropriate for explaining his theory. There may be some disagreement about how to interpret Einstein's theories, but these disagreements are not so pervasive and intractable that a dialogue and debate format is required. Thus, the Appendix is an expository essay that lays out the fundamentals of Einstein's theory and contemporary physical cosmology and discusses the metaphysical implications of these ideas.

This book aims to present original theories we have developed and yet at the same time be accessible to beginners in the field. The new theories advanced will make it of interest to philosophy professors and graduate students, and its accessibility to beginners will make it suitable for use by the general public and in undergraduate courses on metaphysics, philosophy of science and the introduction to philosophy. In an effort to make the book as accessible as possible, we have placed a glossary, study questions and suggestions for further reading at the end of each dialogue and at the end of the Appendix. We have provided the diving board, but once the reader takes the dive into the abyss of metaphysical complexities, the reader's own reasoning powers will be the only guide. The rational struggle for metaphysical truth is a struggle unto death and perhaps the one absolutely certain metaphysical thesis is that death – after allowing us to skirmish for a brief while – will proclaim its silent victory. But the time is not yet, so let us enter the skirmish.

NOTES

1 W. L. Craig and Q. Smith (1993) *Theism, Atheism and Big Bang Cosmology*, Oxford: Oxford University Press.
2 Stephen Hawking (1988) *A Brief History of Time*, New York: Bantam Books.

Part I

The finite and the infinite

Quentin Smith

Dialogue 1

The beginning of time

ALICE: A philosophy major
PHIL: A philosophy major
SOPHIA: A philosophy professor
IVAN: A philosophy professor

Outside of the Lonesome Hut cabin, near the top of Mount Washington in the White Mountains in New Hampshire. Alice, Phil, Sophia and Ivan sit on a promontory, watching the stars emerge as dusk deepens. After a long silence, Alice begins to speak.

ALICE: Do you think that the universe was always here, that time stretches back earlier and earlier into the past, without any beginning at all? Or did everything have a beginning? Did time begin?

PHIL: Time begin? How could *time* begin? Everything that begins, begins in time, so time itself cannot begin.

ALICE: Explain that. Why cannot time begin?

PHIL: If something begins, that means there was an earlier time at which the thing did not exist and a later time at which the thing exists. So if time itself began, that would imply there was an earlier time at which time did not exist and a later time at which time exists. In short, it implies there is a time earlier than the earliest time. It is clear that is a contradiction. So past time must be without a beginning. It is infinite.

SOPHIA: Let me interject here. I understand what you are saying, Phil, but there is a different way of viewing the matter. Time does not begin in the same way that things in time begin. Time begins if there is a first moment, a moment before which there are no other moments. There are two senses of "Something begins." In one sense it means "There was a time at which something did not exist and a later time at which it does exist" and in another sense it means "Something is the earliest moment of time, so that all other moments are later than it." Time "begins" in the second sense.

PHIL: That is hard to conceive. Is there any reason to think that time did begin?

SOPHIA: Some scientists today believe that time did begin. They believe that time began about 15 billion years ago with the so-called "big bang," an explosion of matter, energy and space out of nothingness. At the first time, there existed only an extremely tiny but very dense speck of matter. This speck of matter was so tiny it was much smaller even than an atom. This speck of matter instantaneously exploded in the powerful explosion called the big bang. As this matter exploded, the space containing it began to expand, much like the surface of an expanding balloon. This expansion is still going on today; the universe is growing larger in volume at every moment.

IVAN: Sophia is right. The physicist Stephen Hawking says that time began with the beginning of the big bang explosion. He says that to ask what occurred *before* this explosion is like asking what is north of the North Pole. There is nothing that is north of the North Pole, and there is nothing that occurred before the big bang.

PHIL: I have some difficulty in understanding this idea, and am not quite sure I really can believe it, despite the fact that Stephen Hawking and some other scientists believe it is true. Suppose I am somehow located at the moment of the big bang. I can then conceive of a moment before this moment and if I can do this, that seems to suggest there is an earlier moment.

ALICE: I can see the fallacy in that argument. Surely you could *conceive* of an earlier time, but that does not mean there *was* an earlier time. In fact, your concept will not refer to anything, since there was no earlier moment for it to refer to. You are at the big bang, trying to think of an earlier time, but in fact there is in reality nothing corresponding to your thoughts.

PHIL: But I think your own remarks imply there must have been an earlier time. Consider the statement, *there was nothing before the big bang*. That implies that there was a time before the big bang, a time at which there was nothing.

IVAN: Let me give a precise and logical formulation of your argument, Phil, and see what Sophia has to say about it. I don't agree with the argument myself, but I am interested to see how Sophia responds to it. The argument goes like this. Either there was something occurring before the big bang or there was nothing occurring before the big bang. In either case, there is an earlier time at which *there either was something or nothing occurring*. The contradiction also appears when we say "There was no time before the big bang" for "there was" and "before" refer to the past time that preceded the big bang.

SOPHIA: I think you are again confusing talk about time with talk about things in time. "There was no time before the big bang" does not mean "At the time before the big bang, there was no time." Suppose we are all located at the big bang. We could then take "There was no time before the big bang" to convey that "It is now true that *time Is and Will Be* but it is now false that *time Was*."

IVAN: Sophia, let me elaborate on your point, which seems on the right track. If we are located at the big bang, then we should say that the notion of "there was . . ." has no application at all. "There now is . . ." refers to something and "there will be . . ." refers to something, but "there was . . ." does not. Likewise, we should say that the notion of "before the big bang" has no application at all; only the expressions "simultaneously with the big bang" and "later than the big bang" have application.

PHIL: That is clearer. I think I am beginning to see. Maybe your point can be put metaphorically. If we are at the big bang and try to turn our thoughts towards the past, we encounter an invisible wall of nothingness. There is nothing at all there.

IVAN: Sophia, let me go back to something you said earlier. You stated that "Time begins" means "There is a first moment of time." I do not think that is a very good definition of "Time begins," since there is a sense in which time begins even though time lacks an earliest moment. Time began 15 billion years ago, but for every moment of time, there is an earlier moment.

SOPHIA: Explain exactly what you are getting at, Ivan.

IVAN: Take the first hour of time. Is that the first time? No, since there is a shorter interval of time that elapses before the first hour elapses. There are sixty minutes in an hour and the first of these minutes must elapse before the whole hour can elapse. Moreover, before the first minute elapses, the first second elapses, and before the first second, the first one-tenth of a second, and so on infinitely. Since there is no shortest temporal interval, there is no interval of time that can be identified as the first interval of time. For any interval of time that elapses during the first hour, there is a shorter interval that elapses first. Thus, there is no "first moment of time" in the sense of an interval of time that precedes every other interval.

SOPHIA: But if time begins, there is an earliest interval of time of *each length*. For example, there is a first second, a first hour, a first year and so on. That is what I meant by saying that "Time begins" means "There is a first moment of time."

IVAN: But Sophia, even that is not quite right. It is right only if future time is endless. But it is wrong if time comes to an end. Suppose time begins, and ends 30 billion years later. If so, there is a first second and a first hour and a first year, but there is no first interval of the length *40 billion years*, since no intervals of this length exist. So in this case your statement that "There is an earliest interval of time each length" would be false.

SOPHIA: I see your point, Ivan. What I should say, to be strictly accurate, is that "Time begins" means there is a first interval of time of *some* length (but not necessarily of *every* length). Time begins, for example, if there is a first second, even if there is no first interval of the length *40 billion years*.

IVAN: All right, I agree that is a good definition of "Time begins."

ALICE: I am not entirely sure about everything that you are both agreeing upon, Ivan and Sophia. Ivan, you said that there is no earliest interval of time, since for every actual interval, there is a shorter interval that elapses first. This implies that there are an infinite number of shorter and shorter intervals. But why should we think that time is infinitely divisible? Why can't there be a shortest interval, say an interval that lasts for one-millionth of a second? If there were, then the first interval of that length would indeed be the first interval of time. There would be no interval of any length that elapsed before that interval elapsed.

IVAN: The view that time has a shortest interval that is not further divisible is the view that time is *discrete*. This means that there is some shortest interval of time, say one-millionth of a second, and that there is no period of time shorter than this. Some physicists think that quantum mechanics implies that there is a shortest interval, an interval that lasts for only 10^{-43} second. This is 1/100000 . . . of a second, where there are forty-three zeros after the 1. This is an extremely short period of time, far shorter than one-trillionth of a second. But I think that physicists are best understood as saying that it is impossible to *observe* any interval of time that is shorter than 10^{-43} second. I think it is a conceptual necessity that for any interval of time, there must be a shorter interval.

ALICE: I do not see why this is a "conceptual necessity," as you call it.

IVAN: It is a conceptual necessity since a temporal interval by its very nature contains successive parts. In order for any interval to elapse, the first half of the interval must elapse before the second half elapses. Thus, any interval must be composed of two shorter intervals. For each of these shorter intervals, the first half of the interval must elapse before the second half. This subdivision continues indefinitely, which shows there can be no shortest interval.

ALICE: Perhaps we can *mentally* divide up every interval, but why should it follow that reality must conform to our mental distinctions? Maybe we can mentally divide the interval of 10^{-43} second into two halves, but in reality this interval does not have any parts.

IVAN: A temporal interval by definition has successive parts.

ALICE: Well, maybe your definitions do not correspond to reality. Maybe we should accept what the physicists say.

SOPHIA: I would like to interrupt here. Both of you seem to think that the only alternatives are that time is ultimately composed of intervals of some shortest length (Alice's view) or that time is composed of intervals that are infinitely divisible (Ivan's view). Actually, there is a third theory of time, which is based on the distinction between *instants* and *intervals*. An interval is any time that has duration, however short it may be. But an instant is a time that has no duration at all. According to the third theory of time, time is ultimately composed of instants.

PHIL: Your distinction between instants and intervals sounds mysterious. Could you tell us more about instants?

SOPHIA: An instant lasts for no time at all. It is shorter than one second, one-millionth of a second, one-trillionth of a second, and so on. No matter how short of a period of time you consider, an instant is briefer than that. In fact, each instant lasts for zero seconds. This is what it means to say that an instant exists "instantaneously." It lasts for no time at all. And yet each interval is composed of instants. Indeed, each interval is composed of an infinite number of instants. And each interval, be it one second long or one year, is composed of the exact same number of instants, an infinite number.

PHIL: That sounds extremely paradoxical. If an interval, say, the first hour during which the universe existed, is composed of instants, each of which lasts for no time at all, how can they add up to a period of time that lasts for an hour? If each instant has zero duration, then zero plus zero equals zero, regardless of how many instants there are.

ALICE: The theory of instants also says that there are an infinite number of instants in every interval. That seems very paradoxical as well. If there are an infinite number of instants in one second and also an infinite number of instants in one year, should not one second last just as long as one year?

SOPHIA: Admittedly, it seems paradoxical to think that an infinite number of zeros could add up to some positive number, such as one hour or ten minutes. But there is a mathematical solution to this problem. First, it is obvious that units can be added together only if they are *countable*. Now units are countable only if they can be placed in a one-to-one correspondence with some or all of the positive numbers (1, 2, 3, 4 . . .). This is the standard mathematical definition of "countable." The important fact is that the number of instants in an interval are too numerous to be placed in one-to-one correspondence with the positive numbers. They are not countable and thus cannot be added together.

ALICE: I do not fully understand your theory. Let me begin with your last point, about instants being uncountable and not being capable of being added together. How does that solve the paradox that zero-duration instants add up to make an interval of some non-zero duration?

SOPHIA: If the notion of addition does not apply to instants, then we dissolve the paradox that "There are an infinite number of zero-duration instants in an interval and yet they *add up* to a positive number." They do not add up at all, since the concept of addition does not apply to them. This may sound like an abstract theory, but it is accepted by many philosophers who are familiar with the mathematics involved.

ALICE: OK, if the instants cannot be added, then we cannot say they "add up to some non-zero duration." But I do not fully understand why they can't be added together. How can the number of instants in an interval be *too many* to be placed in a one-to-one correspondence with all the positive numbers (1, 2, 3, 4 . . .)? There are an infinite number of positive numbers, since the series 1, 2, 3, 4 . . . goes on forever. So how can the number of instants be greater than infinity?

SOPHIA: The short answer is that there are two different kinds of infinity, *countable infinity* and *uncountable infinity*, and uncountable infinity is a larger number than countable infinity.

ALICE: Unless you explain this more fully, it makes no sense to me.

SOPHIA: The number of instants in an interval is the number of all *real numbers*. A real number is any number that can be expressed in a decimal form. For example, 0.31475939373 . . ., with an infinite number of decimal places, is a real number. The totality of real numbers is greater than the totality of all positive numbers. The totality of real numbers is *uncountably infinite*, but the totality of all positive numbers is *countably infinite*. This means the real numbers cannot be placed in a one-to-one correspondence with the positive numbers. There are too many real numbers. Just think of all the decimal numbers that come between just two positive numbers, such as 1 and 2; there is 1.11111111 . . ., and 1.11111121 . . ., and 1.11111131 . . . and so on. Even though positive numbers and real numbers are both infinitely numerous, the number of all real numbers is a higher infinite number than the number of all positive numbers.

PHIL: How does this theory answer my objection, that if each interval is composed of an infinite number of instants, then each interval must be of the same length? Clearly, intervals are of different lengths. One hour is longer than one minute. So how could they both be composed of the same number of instants?

SOPHIA: This can be seen by analogy with two lines of unequal length. Any two lines are composed of the same number of zero-length spatial points. Look:

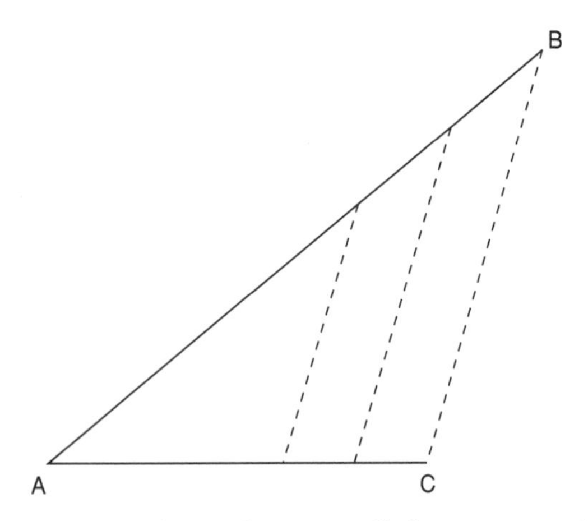

We can draw parallel lines between all the points on A–B and all the points on A–C. This shows that there are just as many points on the longer line A–B as there are on the shorter line A–C. The same holds true for temporal intervals. There are just as many instantaneous temporal points in an hour as there are in a minute.

ALICE: I think I am beginning to get some handle on these ideas. But how does this theory give us a new definition of "Time begins"?

SOPHIA: "Time begins" could mean "There is a first instant of time." There can be no first interval, since for every interval there is a shorter interval that elapses first, but there can be a first instant.

ALICE: You are presupposing that the physicists who say that intervals of 10^{-43} second are the shortest intervals are wrong. If these physicists are correct, then there are no instants at all. This would be because instants would be briefer than the intervals of 10^{-43} second and according to the physicists no time is briefer than these intervals. If these physicists are right, then your definition of the beginning of time as "There is a first instant" is false. The correct definition would be that "There is a first interval of 10^{-43} second."

PHIL: So now we have three definitions of "Time begins." Ivan agreed with the theory that time begins if there is a first interval of some length, say a first second. Alice adopted the theory that intervals are not infinitely divisible, and that there is a shortest interval, the interval of 10^{-43} second. According to this theory, time begins if there is a first interval of the shortest length. Sophia's theory of instants implies that time begins if there is a first instant.

SOPHIA: The theory of instants also allows the possibility that time can begin if there are instants but no first instant.

PHIL: What? How could time begin if there is no first instant?

SOPHIA: If there are instants, then time is *dense*. This means that between any two instants, there is a third instant. In other words, there is no such thing as an instant that immediately precedes a later instant. If we call one instant "1" and a later instant "2," then it is never the case that the instant 2 immediately follows the instant 1. Before the instant 2 and after the instant 1, there is the instant we may call "1.5." But the instant 1.5 is not immediately before 2. After the instant 1.5 and before the instant 2, there is the instant 1.75. And so on. This is clear from the fact that the number of instants in an interval is the number of all real numbers, all numbers that can be expressed by some decimal. Between any two decimals, there is a third decimal.

PHIL: I do not see how this relates to your paradox that the time can begin even if there is no first instant.

SOPHIA: Take a normal hour. It has a first instant and a last instant. For example, the hour from 12 noon to 1 o'clock has a first instant that is exactly 12 noon and a last instant that is exactly 1 o'clock. Now delete the instant that is 12 noon. There is still an hour, since if you subtract one instant from the infinite number of instants that compose the hour, you still have an infinite number of instants and one hour. But this hour has no first instant! Why? Because there is no instant that immediately follows the deleted instant that is 12 noon. Remember, I have just argued that time is dense, that there is no instant that immediately succeeds any other instant.

PHIL: I can see how this relates to the beginning of time. The first hour of time may have no first instant. In that case, time will begin (there will be an earliest hour), but there will be no earliest instant. For each instant that exists, there will be an earlier instant. But I fail to see why we should believe this should be the case if time begins. Suppose time began 15 billion years ago with the big bang explosion. Why should we suppose that the first hour had no first instant?

SOPHIA: The big bang theory implies that if there were a first instant, call it T = 0, the size of the universe would be zero at that instant. Since the universe cannot have zero size, but must have some positive size, however small, there cannot be a first instant T = 0. Rather, the earliest instants that actually exist are those that come after T = 0. But there is no instant that immediately follows T = 0. It cannot be T = 1, since there is an earlier instant, T = 0.5. And that cannot be the earliest instant, since there is an earlier instant, T = 0.25. And so on.

PHIL: We have been discussing theories of the beginning of time. But we have also been discussing theories of the nature of time. It seems we have three theories of the nature of time. Sophia thinks that there is real number of instants in every interval. Alice seemed to like the idea that there is a shortest interval of some length, say the length 10^{-43} second that physicists discuss. And Ivan advanced the theory that time consists only of intervals, with each interval being infinitely divisible into shorter intervals. Which theory is true?

SOPHIA: I think the theory of instants is true. Here is an argument why time should contain instants. The equations of Einstein's Theory of Relativity contain a time variable that ranges over all real numbers. And that requires there to be a real number of zero-duration instants. Since Einstein's theory is confirmed by the evidence, we should believe in the theory of instants it implies.

ALICE: But Ivan pointed out that some physicists said there is a shortest interval, 10^{-43} second.

IVAN: That's right. Quantum mechanics implies there is a shortest interval, 10^{-43} second, but Einstein's Theory of Relativity implies there is no shortest interval and that there are instants. These two theories are incompatible, and yet both are confirmed by different observational evidence. Physicists haven't yet figured out how to combine the two theories into one unified theory. But I don't think we need rely on physics. My theory that time contains intervals only, and that each interval is composed of two shorter intervals, is the most intuitively plausible and has the fewest conceptual difficulties, and so I think we should accept it.

PHIL: Well, I doubt Sophia or Alice would agree with that. So it appears our debate is inconclusive. There is no certain answer available yet about the nature of time and no certain answer about how we should define the beginning of time.

SOPHIA: You are being too pessimistic, Phil. We all agree that time is composed of intervals. And we all agree that "Time begins" means *at least* that there is a first interval of some length, for example, a first second or a first hour. We disagree about whether the intervals are infinitely divisible and whether or not intervals are ultimately composed of zero-duration instants. And we also disagree about the additional meanings that "Time begins" may have, for example, whether or not it also means there is a first instant or a first interval of the shortest length, 10^{-43} second. But the basic core of agreement we share is enough for us to have a common conception of time.

GLOSSARY OF TERMS

Continuous time Time is continuous if it is both dense and the number of instants in any interval is the number of all real numbers. Real numbers are the numbers expressed by decimals, such as 0.3145567 . . .

Countable infinity An infinite series is countable if its members can be matched one-to-one with all of the positive whole numbers (1, 2, 3, 4 . . .).

Dense time Time is dense if it is composed of instants and if between any two instants, there is a third instant.

Discrete time Time is discrete if there are no instants and there are intervals of a shortest length, for example, 10^{-43} second.

Instant A time that has zero duration. There are uncountably infinitely many instants in every interval of time, regardless of how long or short the interval.

Interval A time that has some duration. An interval can be of any length, from one year or longer to one-millionth of a second or less.

Uncountable infinity An infinite series is uncountable if its members are too numerous to be matched one-to-one with all of the positive whole numbers. The infinite series of real numbers is uncountably infinite.

STUDY QUESTIONS

1 Why is this sentence potentially misleading: "Before the earliest time, there was nothing at all occurring"?

2 Explain three possible ways to define "Time begins."
3 What is the one certain definition of "Time begins" upon which all the philosophers can agree?
4 Is *3 o'clock* an instant or an interval of time?
5 Explain how it could be true that "Time ends" and yet that there is no last interval of time.
6 Explain how it could be true that "Time ends" and yet that there is no last instant of time.

FURTHER READING

A selection of advanced and introductory readings on the topics discussed in this dialogue are listed in what follows. An asterisk is placed before the readings that are more advanced and technical in nature.

*Grünbaum, Adolf (1968) *Modern Science and Zeno's Paradoxes*, London: George Allen and Unwin Ltd.
This book contains one of the best technical discussions of the relation between instants and intervals and how an infinite number of zero-duration instants can compose an interval of positive duration.
Newton-Smith, W. H. (1981) *The Structure of Time*, London: Routledge.
Newton-Smith's book is a mostly clear introduction to various topics in the philosophy of time and contains some good arguments for the possibility that time begins.
Salmon, Wesley (1980) *Space, Time and Motion*, Minneapolis: University of Minnesota Press.
This is a clear introduction to various topics mentioned in its title, including the theory of instants and intervals that is treated more technically in Grünbaum's book.
Smith, Quentin (1985) "On the Beginning of Time," *Noûs* 19: 579–84.
This paper contains several arguments that it is possible for time to begin.
*—— (1989) "A New Typology of Temporal and Atemporal Permanence," *Noûs* 23: 307–30.
In this essay several different senses in which something can be "permanent" or "begin and cease to exist" are classified.

Dialogue 2

The infinity of past and future time

ALICE: Is it certain that time began with the big bang?

SOPHIA: No, it is not known for sure. Most physicists think there is some probability that it is true, but it is not absolutely certain.

ALICE: So it is possible that past time might instead be infinite?

SOPHIA: Yes. It might be the case that time has no beginning. Some physicists think that if time has no beginning, then the big bang that occurred 15 billion years ago was the result of a prior phase of contraction. The universe expands from a big bang, then contracts, then expands again from another big bang, then contracts, and so on an infinite number of times. If time is infinite, there may have been an infinite number of expansions and contractions of the universe before the present phase of expansion that began 15 billion years ago.

ALICE: How should we understand this theory in relation to our earlier discussion of instants, intervals and the beginning of time?

SOPHIA: It is important to note that when we say "The past is infinite" we are talking about *intervals* of time, not instants. If time began one hour ago, there are an infinite number of past instants. The number of past instants is infinite regardless of whether or not time had a beginning. But when we say things like "The universe has an infinite past," we normally mean that there are an infinite number of intervals earlier than the present interval.

IVAN: I agree with what you are saying Sophia, but you need to add a qualification. You need to say that there are an infinite number of intervals *of the same length* before the present interval. For example, there are an infinite number of years before the present year, an infinite number of seconds before the present second, and so on. Otherwise, someone might say that there are an infinite number of intervals before the present interval and yet that time began. For example, before the present year, there was a present hour, a present second, a present one-millionth of a second, and so on infinitely. But in this case, time may still have begun.

PHIL: Sophia and Ivan, your account seems fairly clear on the surface. But when I think more closely about the idea that the past is infinite, it no longer seems to make sense. What exactly does it mean, for example, to say that an infinite number of years have elapsed before the present year?

SOPHIA: Let us number the years in this way. The present year is number 0, last year is –1, the year before that –2, and so on. Then we have a countably infinite number of past years, with a different past year corresponding to a different number in the negative number series:

Past years	Present year
. . . –5 –4 –3 –2 –1	0

PHIL: At what number does the series end?

SOPHIA: It doesn't, it just keeps going on endlessly through all the negative numbers.

PHIL: Let me raise a question that I never had a chance to raise when we were discussing the infinity of instants and intervals. I am not clear about the meaning of the phrase "infinite number." If there is an infinite number of past times, must not there be a past time that corresponds to the highest number, the number *infinity*, that comes at the end of the series?

SOPHIA: No, infinity is not a number at the end of the series. Rather it is the number of all the numbers in the series. The series itself has no end, it just keeps going.

PHIL: Well, if time is like that, how could a year that once was present ever recede an *infinite* distance into the past? A year that once was present can recede a finite distance into the past. But it cannot recede an infinite distance, since that would require traversing an infinite temporal series. It is the very nature of an infinite series that it has no end and therefore that it cannot be traversed. For every member of the series that is traversed, there is still one more member that needs to be traversed. So how could a year that once was present recede an infinite number of years into the past?

ALICE: Are you suggesting that the infinitude of the past implies that there is some distant year that is separated by an infinite number of years from the present year?

PHIL: Yes. If the past is infinite, then there must be some year that is separated from the present by an infinite number of years. If every past year is separated from the present year by only a finite number of years, then past time is finite. For example, if there is a year that is 60 billion years in the past, that implies that the past is 60 billion plus one years old. And the same holds for any other year that is separated from the present year by only a finite number of years.

SOPHIA: Phil, you are forgetting that there are an infinite number of years in the past. Each year is only finitely distant from the present year, but there are an infinite number of finitely distant years. That is all it means to say the past is infinite.

IVAN: It is like the negative number series. Every negative number, be it minus 65 or minus one trillion, is separated from zero by a finite number of negative numbers, but it is nonetheless truth that there are an infinite number of negative numbers.

PHIL: OK. I can understand that. But I am still not entirely satisfied with the idea of an infinite past. If the past were infinite, then new years could not be added to it. If the past were finite, say 15 billion years old, then new years could be added to it. Suppose we assign a negative number to each past year. If there are only 15 billion past years and the present year becomes past, there is a new number to assign to it, namely, –15,000,000,001 (minus 15 billion and one). But if the past years were infinite, then all finite numbers would be assigned to the past years and there would be no finite number left over to assign to the present year when it becomes past. So if past time really is infinite, the present year could never become past!

ALICE: That seems like a paradox.

PHIL: It is more than that; it is a refutation of the idea that the past is infinite. Essentially, the problem is this. If the past is finite, then the past is incomplete in the sense that there is always room for one more year to be added to it. If there are 15 billion past years, you can also add one more, a 15 billionth-plus-one-year. But if the past is infinite, then it is complete in the sense that there is no room for any more additions. You cannot add a 15 billionth-plus-one-year, since that year already belongs to it. And the same holds for any other year with a finite number attached to it.

ALICE: It seems like Phil has a point, Sophia. How do you respond to that?

SOPHIA: Let us consider first this question: what happens when you subtract infinity from infinity?

PHIL: You get zero of course.

SOPHIA: No. You get *infinity.*

PHIL: That's crazy! How can you subtract infinity from infinity and still get infinity?

SOPHIA: Consider the series of all negative numbers:

. . . –7 –6 –5 –4 –3 –2 –1

Suppose you subtract or remove all the odd numbers (–1, –3, –5 . . .). There are an infinite number of odd numbers, so you have subtracted an infinite number of numbers. But how many numbers are left? Obviously an infinite number, since there are an infinite number of even numbers.

ALICE: I still don't see how that resolves Phil's paradox.

SOPHIA: It can be resolved in several ways. One way is to renumber all currently past events by negative even numbers only, so that we have

Past years	Present year
. . . –8 –6 –4 –2	0

Since there are an infinite number of negative even numbers (the same number as all negative numbers), this implies there are an infinite number of past years. But this allows us to use the odd negative numbers, –1, –3, –5 and so on, to number the new years that become past.

ALICE: How would that work?

SOPHIA: Suppose the present year 0 becomes past. Then we make the new present year number –1:

Past years	Present year
. . . –8 –6 –4 –2 0	–1

When –1 becomes past, we have to assign the next odd negative number, –3, to the new present year. When that becomes past, we assign –5 to the next present year, and so on without end.

PHIL: I think I understand now.

ALICE: Don't be too quick about saying this really makes sense,

Phil. For with these additions, we no longer have an infinite number of past events, but a greater number, infinity plus three new years, the years 0, –1 and –3. So it is not true that the past is still infinite!

SOPHIA: The number of all whole numbers is called "aleph-zero," $\aleph_0$, by mathematicians. "Aleph-zero" is the name of countable infinity. An interesting peculiarity of aleph-zero is that if you add one to it, you still get aleph-zero. If you add one to the number eight, you get nine, but if you add one to aleph-zero, you still have aleph-zero. $\aleph_0 + 1 = \aleph_0$. This is not all that surprising, since infinity is very different from any finite number. How can you increase infinity by adding a finite number to it? Clearly, you cannot. Infinity plus any finite number is just infinity. So if we add three new years to an aleph-zero number of years, we still have aleph-zero years.

ALICE: While you were explaining that a new question occurred to me. *When* would the number of past years become infinite? There must have been some time in the past when the number of past years stopped being finite and became infinite. But I cannot conceive how this could happen. For example, suppose 1960 was the year that made the past become infinite. How could that happen? If in 1959 the number of past years is finite, how could the addition of one more year make the number infinite?

SOPHIA: There never was a time when the past became infinite. It always was infinite.

PHIL: What? That's absurd! How could the past be infinite without first having been merely finite at some point?

SOPHIA: It's just impossible for the past to become infinite by the passing of a year, say 1960. If a finite number of years have passed, one more year is not going to make a difference. If you add one year to a finite number of years, you still get some finite number. Even if you add 10 billion years to a finite number of years, you still get a finite number of years, just a larger finite number.

ALICE: So you are saying that no matter how many years you go back into the past, you will never reach a year when there was only a finite number of years before that year. If you go back to 100,000BC, say, you will find that there are the same number of past years then as there are now, namely, the number you call aleph-zero.

PHIL: But that can't be right. Obviously, there are more past years *now*. There are 100,000 more past years!

SOPHIA: But you are forgetting about the nature of infinity. If you add 100,000 to aleph-zero, you still get aleph-zero.

PHIL: Then I have a challenge for you. Suppose I go back an *infinite* number of years into the past. What will happen then? What would I find?

ALICE: I know the answer. Since infinity subtracted from infinity is still infinity, you would still find the past to be infinite if you went back an infinite number of years into the past.

SOPHIA: No, Alice, that is not right. There is a crucial difference between *subtracting* an infinite number of years from the past and *going back an infinite number of years into the past*. (By "going back into the past" I mean thinking yourself back through time.) If you go back an infinite number of years into the past, you go back through the years in the reverse order in which they occurred and you will traverse all of past time.

IVAN: I see what Sophia is getting at. The point is actually quite simple. If you travel back into the past, you travel back over all the years in the order in which they exist. Suppose all past years are numbered 0, –2, –4, –6, –8 . . . You do not travel by jumping from 0 to –4 to –8; you go from 0 to –2 to –4 to –6 and so on. But if you travel back in this way, the only way to go back an infinite number of years is to traverse all the past years.

ALICE: It seems that if you traverse all the years, you must in some sense reach a final year in your journey, and this would be the earliest year in time.

IVAN: No! There is no earliest time if the past is infinite. You traverse all the years, but you do not reach a final year since there is no final year in your journey. You would go back forever into the past, without ever meeting a year that was not preceded by an earlier year.

ALICE: I think I am beginning to understand. But what about future time? Is there any difference between the future being infinite and the past being infinite?

IVAN: Not really. Infinite future time is best modeled on the series of all positive numbers, with the present year being zero and future years being numbered as 1, 2, 3 . . . If the past and future are both infinite, then they will correspond to the series of all negative and positive numbers.

Past years	Present year	Future years
... –5 –4 –3 –2 –1	0	1 2 3 4 5 ...

PHIL: Is there any scientific evidence that the future will be infinite?

IVAN: The issue is undecided as of yet. The current observational evidence is consistent both with the hypothesis that the future is infinite and with the hypothesis that the future is finite.

PHIL: What are the two different possible futures of the universe?

IVAN: If the future is finite, then the universe will eventually stop expanding and begin contracting. It will keep contracting until it shrinks down to nothing and ends with a final "big crunch," as scientists call it. This will be the end of time. If the future is infinite, the universe will expand for an infinite amount of time. All the stars will burn out and all matter will eventually break down into radiation, which will become cooler and cooler. The universe will become dark and cold, and will be like an endlessly expanding tomb.

PHIL: So if the future is infinite, change will never come to a stop. The universe will always be expanding.

SOPHIA: In a sense that is right, Phil. But Ivan has left out an intriguing possibility. There is something very interesting about the future of our universe that makes it different from the past. We have agreed that if the past is infinite, then there does not need to be any past year that is separated from the present year by an infinite number of years. Each past year could be finitely distant from the present year. This, of course, could also be the case with the future, if all that happens is that the universe keeps expanding forever. But if the universe *comes to rest after it has expanded for an infinite time*, then future time will possess a very different type of infinitude than the past.

PHIL: You have lost me. What is this "different type of infinitude"?

SOPHIA: The infinitude of the past is commonly thought to have the type called omega-star, after the Greek letter *omega* with a star, ω^*. This is the order type of the negative numbers (. . . –3, –2, –1). The infinitude of the future is commonly thought to have the order type called omega, ω. This is the order type of the positive numbers (1, 2, 3 . . .). But it is possible for the past and future to have different order types as well. Let us concentrate on the future, since the scientific evidence is consistent with the order type of the infinite future being $\omega + \omega$, rather than simply ω.

PHIL: "Omega plus omega"? What does that mean?

SOPHIA: It means you run through *two* infinite series, each one with the order type of the positive numbers. It can be represented mathematically if we use the odd positive numbers for the first infinite series and the even positive numbers for the second infinite series.

The present	The future	
0	1 3 5 7 9 . . .	2 4 6 8 10 . . .

After the infinite time-series corresponding to 1, 3, 5, 7, 9 . . . has elapsed, another infinite time-series begins, that corresponding to 2, 4, 6, 8, 10 . . .

IVAN: Sophia, this idea of yours seems rather unorthodox. What makes you think that the scientific evidence suggests the future may be like this?

SOPHIA: The expansion of the universe could last for an infinite number of years, with its rate of expansion steadily getting slower and slower. After an infinite number of years elapse, the universe will be at rest and remain in an absolutely motionless stillness for another infinite number of years. The expansion of the universe takes place during the infinite number of years corresponding to 1, 3, 5, 7, 9 . . . After this infinite series has elapsed, the universe remains at rest for another infinite series of years, the years corresponding to 2, 4, 6, 8, 10 . . .

IVAN: I do not think that the scientific evidence suggests that this will occur. Moreover, your theory of two infinite time-series seems flawed on purely conceptual grounds. We have just agreed that in an infinite series of years, no year is separated from the present year by an infinite number of intervening years. For each future year, there is only a finite number of years between it and the present year. But according to your theory, there is an infinite number of years separating each year in the second infinite series from the present year. For example, the year corresponding to 2 in the second series is separated by an infinite number of years from the year corresponding to 0. It is separated from the year corresponding to 0 by all the years corresponding to 1, 3, 5, 7, 9 . . .

SOPHIA: But you are presupposing that the future has the order type ω (1, 2, 3 . . .). If the future has the order type $\omega + \omega$ (1, 3, 5 . . ., 2, 4, 6 . . .), then some future years are separated from the present year by an infinite number of intervening years.

IVAN: How could the second infinite series begin *after* the first infinite series? The first series has no end; for every year in the first series, there is a later year. Consequently, it is impossible for this series to end and for a second series to begin.

SOPHIA: Ivan, you are confusing categories of finite time with categories of infinite time. The first infinite series *does* come to an end. It does not end after a finite amount of time, but it does end after an infinite amount of time. An infinite amount of time elapses and the series is over, and the next series begins.

IVAN: Look, for each year in the first series, there is a later year. So there is no year in the first series at which you are going to find yourself at the end of the first series. To find yourself at the end of the first series, you have to find yourself at the last year of the series, a year after which there are no more years in the series. But since the series is infinite, there is no such year.

SOPHIA: You are presupposing a sense of "end" that applies only to finite temporal series. If a finite temporal series ends, then there is a last year. But there is no last year if an infinite series ends; rather, it ends in the sense that "Every year in the infinite series has elapsed." The matter can be put in this way. It is true of both finite and infinite series that they "end" in the sense that "Every year in the series has elapsed." But it is true only of a finite series that it "ends" in the sense that "There is last year in the series and it has elapsed."

IVAN: I have a decisive refutation of your argument, Sophia. Suppose I were situated in the first year of the second infinite series, the year corresponding to the number 2. If I were at that year, I would know that some year was *one year earlier than the year corresponding to* 2. But according to your theory, no year could be one year earlier than the year corresponding to the number 2. Your theory implies there is no last year in the series 1, 3, 5, 7, 9 . . ., and therefore that there is no year that immediately precedes the first year in the series 2, 4, 6, 8, 10 . . .

SOPHIA: Ivan, you are simply presupposing that time cannot have the order type $\omega + \omega$. If it does have this order type, then there is no year that immediately precedes the year corresponding to 2.

If you were at the year corresponding to 2, and you thought "There is a year that is *one year earlier than the present year*," then your thought would not refer to anything in reality. The correct way to grasp the preceding time would be to think "For any year that is earlier than the present year, there are an infinite number of years intervening between it and the present year."

IVAN: That is patent nonsense, Sophia.

SOPHIA: It is nonsense only if you assume that time cannot have the order type $\omega + \omega$. But all your arguments simply *assume* that time cannot have this order type; they do not *prove* that time cannot have this order type.

ALICE: Sophia, it seems to me that Ivan is correct. I cannot imagine myself being located at the year corresponding to 2 and believing that there is no immediately preceding year. It seems to me that you are trying to apply a mathematical theory of the infinite to time, and what makes sense in the mathematical realm need not make sense when applied to time.

SOPHIA: Mathematics is based on logic and thus is the very bedrock of what makes sense. If the application of mathematics to time does not seem to make sense to you, this is because you are not following logic but are presupposing your common-sense beliefs about time. If these common-sense beliefs conflict with mathematics and logic, these common-sense beliefs ought to be rejected. *Our understanding of time is based on our understanding of the nature of numbers*, as all our discussions up to this point have shown. If you can order numbers with the order type $\omega + \omega$, then you can also order years with this order type. There is no relevant difference between numbers and years that prevents years from having this order type. Nothing that you or Ivan has said has proven there is any relevant difference.

PHIL: It seems like we have reached an impasse here. Maybe we should proceed to another subject.

IVAN: I agree. In fact, there is a topic very closely related to Sophia's theory of the future. Sophia's theory implies that after the universe's expansion and all other types of change have ceased, time will still continue to pass. I have raised a number of objections to Sophia's theory, but there is another objection that could also be made. This is the objection that *time requires change*. Therefore, time cannot continue to pass after all change has stopped.

SOPHIA: The issue of the relationship between time and change is a very complex one, and we haven't really touched on this issue yet. This issue involves the distinction between the *relational* and *substantival* theories of time. I think we should discuss this distinction next.

GLOSSARY OF TERMS

Aleph-zero This is the number of all whole numbers (. . . –3, –2, –1, 0, 1, 2, 3 . . .). This number is a countable infinity. The number of all whole numbers is the same number, aleph-zero, as the number of all positive whole numbers (1, 2, 3 . . .). This is proven by the fact that all whole numbers can be paired one-to-one with all the positive whole numbers.

The order type ω This is the order type of the positive numbers (1, 2, 3 . . .). Future time may have this order type.

The order type $\omega + \omega$ This is the order type exemplified by the sequence of positive odd numbers followed by the sequence of positive even numbers (1, 3, 5 . . ., 2, 4, 6 . . .). It can be argued that it is possible for the future to have this order type.

The order type ω^* This is the order type of the negative numbers (. . . –3, –2, –1). Past time may have this order type.

STUDY QUESTIONS

1. If the past is infinite, must there be a past year that is separated from the present year by an infinite number of intervening years?
2. If there is no earliest year of time, is the number of past years countably infinite or uncountably infinite?
3. If time will end in 30 billion years, does this imply that the total number of years that make up time is finite?
4. Do you agree with Sophia or Ivan about whether or not future time can have the order type $\omega + \omega$?
5. "It is possible for infinite past time to have the order type $\omega^* + \omega^*$." Explain what this sentence means.

FURTHER READING

*Eells, Ellery (1988) "Quentin Smith on Infinity and the Past," *Infinity and the Past* 55: 453–5.

GARDNER HARVEY LIBRARY
Miami University-Middletown
Middletown, Ohio 45042

This brief article presents a novel solution to a paradox that is involved in the idea that the past is infinite.

Craig, William Lane (1979) *The Kalam Cosmological Argument*, New York: Harper and Row.

Craig's book is the most important and influential work to appear that defends the idea that past time cannot be infinite.

—— (1991) "Time and Infinity," *International Philosophical Quarterly* 31: 387–410.

This article contains Craig's response to Smith's criticism (in the article listed below, "Infinity and the Past") of Craig's book *The Kalam Cosmological Argument.*

—— (1993) "Smith on the Finitude of the Past," *International Philosophical Quarterly* 33: 225–31.

In this article Craig responds to Smith's criticisms in "Reply to Craig: The Possible Infinitude of the Past" (listed below).

Craig, William Lane and Smith, Quentin (1993) *Theism, Atheism and Big Bang Cosmology*, Oxford: Clarendon Press.

In Part I of this book, some of the critical exchanges between Craig and Smith on whether the past can be infinite are presented.

Smith, Quentin (1987) "Infinity and the Past," *Philosophy of Science* 54: 63–75.

This article criticizes Craig's and other philosophers' arguments that the past cannot be infinite.

*—— (1989) "A New Typology of Temporal and Atemporal Permanence," *Noûs* 23: 307–30.

This article includes an argument that time can consist of an infinite number of successive time-series, with each time-series itself including its own infinite past and infinite future.

—— (1993) "Reply to Craig: The Possible Infinitude of the Past," *International Philosophical Quarterly* 33: 109–15.

This essay is a criticism of Craig's argument that the past is necessarily finite that is presented in his above-mentioned article "Time and Infinity."

Dialogue 3

The relational and substantival theories of time

SOPHIA: The issue about time and change does not concern merely the case where the universe expands for an infinite time. It also concerns the case where the universe stops expanding after a finite time, starts to contract, and finally comes to an end in the big crunch. If time does not require change, time could continue flowing on after the universe has contracted to a small point and vanished. There could be empty time elapsing after the universe has ceased to exist. Let us examine this general question for starters: if all change comes to a stop at some time in the future, must time come to an end?

PHIL: What would it be like for time to come to an end?

ALICE: I think it would be like a movie that suddenly freezes at one frame and becomes a still picture. The characters would be caught forever making their final gesture, thinking their last thought, with an expression of boredom or incredulity frozen on their countenances.

PHIL: But that example suggests that time would go on if change stops. You said that people would be caught forever making their final gesture. That implies that time goes on forever, with people frozen fast in an unchanging universe.

IVAN: That's right Phil. The example is not appropriate. A better example would be a movie suddenly freezing at one frame but the people not *remaining* frozen in their last gestures. Rather, the movie and the whole movie theater, indeed, the whole universe, would vanish into thin air. It would not be the case that *after* the end of change, nothing happens, since there is no "after." The last change is the last time.

ALICE: That is an interesting statement, that "The last change *is* the last time." That suggests that a time is nothing but a change.

IVAN: In a sense yes. But we can be more exact. Consider any time, say right now, 9:00 p.m., April 30, 1991. What is this time? Suppose I said to you, go out and find the time called "9:00 p.m., April 30, 1991." What would you find?

ALICE: [*waving her hand*]: Here, that is the time, my moving my hand.

IVAN: That's partly right. The time called "9:00 p.m., April 30, 1991" partly consists of your moving your hand. But strictly speaking, a time is not just one event or occurrence, but a collection of all the events that are simultaneous with any given event. The time called "9:00 p.m., April 30, 1991" is the collection of all the events simultaneous with your waving your hand (including of course your handwave itself).

PHIL: Is it nothing more than that?

IVAN: The description "9:00 p.m., April 30, 1991" implies that it is a collection of simultaneous events that stand in relation to some other collection of simultaneous events. It stands in relation to the collection of simultaneous events that includes the birth of Christ, or at least that is traditionally understood as the birth of Christ (some people think Jesus was born in 3AD). "9:00 p.m., April 30, 1991" means that the collection of events is 1,990 years, 3 months, 29 days and 21 hours later than the collection of events that includes Christ's birth.

ALICE: What then is time, given that a time is a collection of simultaneous events?

IVAN: A time-series is all the collections of simultaneous events, in the order in which they occur. There is one collection for 9:00 p.m., April 30, 1991, a later collection for 9:01 p.m., April 30, 1991 and so on. If time has an infinite future, that implies there are an infinite number of these collections after 9:00 p.m., April 30, 1991; but if time ends, there are only a finite number of these later collections.

PHIL: I can see now why time would stop if all change stopped. If a time is nothing but a collection of simultaneous events, then there can be no time after the last event has occurred. The last event to occur would belong to the very latest collection of simultaneous events. There would be no event after that and therefore no "after" at all; time would have come to an end.

SOPHIA: Ivan, you are giving Phil and Alice a rather one-sided view, since you have only explained to them the "relational theory of time." This is the theory that times are but collections of events related by the relations of being earlier than or later than each other. There is also a different theory, the "substantival theory of time," according to which times are entities or "substances" in their own right, distinct from events.

PHIL: I don't understand. How exactly does the substantival theory differ from the relational theory?

SOPHIA: One difference is that if all change stops, time need not come to an end; it could keep on going, flowing on in a universe that has frozen fast.

IVAN: Sophia, I have always found the so-called substantival theory of time rather mysterious. What could a time be if not a collection of events? How could we come across a time?

SOPHIA: Let us call times (according to the substantival theory) *moments*. Now obviously we cannot see moments with our eyes or hear them. But we are nonetheless constantly aware of them. We are always aware of time passing, of the passing moments (except perhaps when we are terribly absorbed in some idea).

IVAN: Sophia, that is a faulty inference. Certainly we are aware of time passing, but you cannot infer from that that we are aware of *moments* (in your sense) passing by. Rather, our awareness of time passing is nothing but our awareness of changes occurring.

SOPHIA: Imagine you are totally becalmed on a sailboat on a cloudless and windless day. The sea seems perfectly still, and no changes at all are appearing to you. And yet you are aware that time is passing. What you are sensing is that moments are passing by.

IVAN: But you are still aware of changes, going on in your mind. Your thoughts and emotions are changing.

SOPHIA: They may be changing, but I am not aware of them; rather, my whole attention is focused outwards, absorbed entirely in the apparently changeless scene and the passing of moments.

IVAN: I think you must be aware at least marginally of your changing mental states.

SOPHIA: No, I am completely absorbed in the external world.

IVAN: I think that is dubious. I certainly am not aware of any such "passing moments." In any case, you would surely admit that the passage of moments is not something that can be "observed" by normal means, the way we can observe somebody running past us or observe a star through a telescope. Everybody would agree that we can observe such things. But at best it is a highly controversial claim that your "moments" can be observed or experienced in any direct way at all. To rest the case for the substantival theory of time entirely upon the claim that you "experience" passing moments gives the theory a shaky foundation.

PHIL: Is there any other argument for the substantival theory besides your argument, Sophia, about experiencing time when you are not experiencing change?

SOPHIA: Yes. I have an argument based on the truth about a certain possibility. Isn't it true that *nothing might have happened now*? Many things are happening now, but isn't it possible nothing whatsoever might have been occurring now? I think it *is* possible, which shows that the time we identify as "now" is something different than the collection of all the simultaneous events.

IVAN: I deny that it is possible. If "Nothing might have happened now" makes any sense at all, it means "The present time might not have existed."

SOPHIA: Then consider this slightly different argument. It is possible that *right now*, completely different events than all the events in the current collection of simultaneous events might be occurring instead of the ones that are occurring. For example, it might have been the case that right now, only three explosions of three star-like objects were occurring.

PHIL: So what does that show?

SOPHIA: It shows that the present time cannot be the collection of all the events that are simultaneous with Alice's handwave. If the present time were this collection, it would not exist if these events did not exist and some other events existed instead. Some other time, some other collection of events, would have existed instead, a collection containing the three explosions.

PHIL: I still don't see your point.

SOPHIA: Well, consider what it means to say that "Right *now, at this time,* completely different events might have existed instead of the ones simultaneous with Alice's handwave." "Right now" means *at the present time*, the time that is 9:00 p.m.,

April 30, 1991. If the three star-like explosions might have existed now, at the time that is 9:00 p.m., April 30, 1991, then this time cannot be the collection of all the events simultaneous with Alice's handwave. It must be something different from and independent of this collection. It must be a *moment*.

PHIL: I see what you are getting at. If one and the same time could have been occupied by entirely different events, then it is not dependent for its existence on the events that occupy it. It is some independent item or "substance" in its own right. And that implies that time could go on existing even if all events stopped.

SOPHIA: I think arguments such as this show that there are moments even if we do not directly experience them. The substantival theory of time does not depend for its truth upon the assumption that we perceive or experience these moments. In fact, many people who hold the substantival theory claim we do not experience them, but that we should posit them to explain what we do experience, much as scientists posit electrons and the like to explain what we do experience.

IVAN: Well, if we don't experience moments, then the hypothesis that these moments exist not only becomes unverifiable but also leads to absurd consequences. Consider, for example, the rate of change of things we experience and the "rate of passage" of the moments in your substantival time. If there are moments of time that constitute "the passage of time," these moments must pass at some rate.

PHIL: What exactly is a "rate of change" or a "rate of passage"?

IVAN: A *rate* of change is how fast one quantity changes in comparison with the changes in another quantity. If I say John runs at the rate of one mile per five minutes, that has the following meaning: I observe John running past one mile-marker simultaneously with the position of "12:00" of the hands on my watch. I watch him run past a second mile-marker simultaneously with the position "12:05" of the hands on my watch. I observe two things, John and the watch, and compare the simultaneous changes they undergo.

My argument against Sophia's "substantival theory of time" is this. If there are unobservable moments, then the changes in the watch would measure or represent or correspond to the passage of these moments. For example, "five minutes" would mean that five minute-length moments had passed (and not simply that the hands on the watch had changed their position).

But then an absurd possibility arises. It is possible that on every Tuesday all physical and mental changes speed up to some enormous degree relative to the rate at which moments of time pass (the "moments of time" would continue to pass at their same old rate); but if this happened, we would never notice it. If all physical processes speed up at the same rate, we wouldn't notice this, since we couldn't observe anything to compare them with. We wouldn't observe the "passing moments" and so we couldn't notice that physical processes speed up relative to them.

However, if there are no moments of time, then such an absurdity would not be possible. Physical processes could not "speed up" since there would be nothing relative to which they would speed up.

SOPHIA: I agree that it is possible that such bizarre things could occur. But all sorts of bizarre things are possible. It is possible that the universe began five minutes ago, replete with false memories, false traces of a distant past, etc. But bizarre possibilities of this sort are not *probable* and so we need not worry about them. It is simpler to assume that the rate at which all physical things changes does not "speed up" on some occasions. The idea that we should adopt the simplest hypothesis and rule out bizarre possibilities for which we have no evidence is a rule of scientific reasoning, and we should adopt it in this case.

ALICE: Can I interrupt your argument here? I am not really sure what the point of the dispute is. What difference does it make whether the relational theory of time is true or the substantival theory?

SOPHIA: One difference it makes is whether time will go on after all change ends (if change ever ends – I earlier suggested that it probably will end). Another difference is that if the substantival theory is true, the big bang, which began the universe, need not be the beginning of time. Moments of time could have extended infinitely into the past, prior to the big bang.

IVAN: The issue is also important because it answers the question "What is time?" Time is either a reality separate from what goes on in the universe or it is nothing but the collection of all the physical and mental occurrences in the universe.

ALICE: OK. I'm satisfied. Proceed with your argument.

SOPHIA: Let me present a third argument for the substantival theory of time. This is based on a thought-experiment; I don't mean to suggest that any of this is ever likely to happen, but it is a *possibility* that tells us something about the nature of time.

Imagine a universe in which things literally vanish at regular intervals and then reappear a certain time later. The vanishing things do not enter into some other universe; rather, they literally cease to exist for a period of time. Assume I am an inhabitant of this universe and that I regularly cease to exist and reappear as well. I would be sitting at my desk and then, without noticing it, would suddenly cease to exist, and then reappear at my desk one week later. I wouldn't notice any change until someone walked in and told me that I had ceased to exist, and that it was now a week later. If I am inclined to doubt that I disappeared, I would ask other people, and they would agree that I had ceased to exist, show me newspapers about what happened during the week I was gone, and would point out other changes (e.g., it was snowing heavily one week ago, just before I ceased to exist, but now the snow has melted).

Now let us suppose that in this universe there is an accurate calculation for each object as to how long it will cease to exist and when it will cease to exist. For example, we know that every six months I cease to exist for exactly a week. Furthermore, every object begins to turn completely white five minutes before it ceases to exist, and is pure white at the last instant before it ceases to exist. When it reappears, it is pure white and gradually changes to its normal color over a five-minute period.

Now it normally happens that in this universe just a few objects are missing at a given time, although on some occasions most of the objects are missing. Now comes the crucial element in this story. Let us suppose that our calculations show that at the year 2000, at exactly 4:00 p.m. on July 5, each object in the universe is going to cease to exist. The first object to reappear will be my pen, which has a vanishing time of only twenty minutes.

At exactly five minutes to four on July 5 in the year 2000, everything begins to turn white, just as predicted. At exactly 4:00 p.m. everything is pure white.

Then what? The next thing anybody notices is when they begin to exist again, whether it be one hour later or one week later or some other time. Things reappear as pure white and return to their normal color after five minutes.

It seems clear that the best explanation of this fact is that everything ceased to exist at 4:00 p.m. and for a twenty-minute period there existed only empty time, with no space, matter or persons. After this twenty-minute period, my pen reappeared and other things reappeared later.

Since this is possible, it shows that empty time is possible. But empty time is possible only if time consists of moments, i.e., substances or entities in their own right.

IVAN: It is possible this could happen. It is possible that there could be empty time and that time is a series of moments that lapse independently of whether any events occupy them. But that does not show the substantival theory of time is true in the actual world. In our world, the relational theory of time is true; times are just collections of events.

SOPHIA: I disagree. We are talking about what time essentially is, what belongs necessarily to its very nature. If a certain characteristic of time necessarily belongs to its very nature, then it belongs to its nature in every possible world in which time exists. Thus, if there is some possible world in which time is a series of independent particulars, then time is such a series in every world. My story about a possible world shows us the necessary nature of time.

IVAN: I do not accept that. I believe the nature of time could vary from world to world.

SOPHIA: But if time varied that much from world to world, it would no longer be time. If you became a stone in another possible world, the stone would no longer be you. It would be something else altogether; you would not exist in that world. Likewise, time cannot change its basic nature from world to world. Time cannot change from being a series of independent particulars to being a collection of events. That is tantamount to time ceasing to exist.

IVAN: Maybe the point can be put in this way. There are two types of time, substantival time and relational time. In some possible worlds, there is substantival time and in others there is relational time. There is substantival time in the world you described. But that does not prove there is substantival time in the actual world. In order to know that, we would need some evidence that empty time elapses.

SOPHIA: I think there is a deep disagreement between us here. I believe there is only one kind of time and it has the same essential properties in every world.

ALICE: It looks like we are not going to reach agreement here. But there are some things we can conclude. If the relational theory of time is true, then time comes to an end when change stops. Thus, there could not be time after the universe ceases to exist or after all change stops in the universe. But if the substantival theory of time is true, then time could continue to pass after the universe ceased to exist or after all change ceased.

GLOSSARY OF TERMS

Relational theory of time According to this theory, a time is a collection of all events simultaneous with a given event and thus time cannot exist without change.

Substantival theory of time According to this theory, a time is an individual in its own right, and can exist even if it is not occupied by any events.

STUDY QUESTIONS

1 According to the relational theory of time, is the event consisting of your reading this sentence *a time*?
2 What is a "rate of change"?
3 Suppose the following sentence is true: "John Doe is asleep now, but he might not have been." How can the fact that this sentence is true be used to argue for the substantival theory of time?
4 Do you think it is possible to be aware of time passing, as a phenomenon distinct from physical and mental changes?

FURTHER READING

*Cover, J. A. (1993) "Reference, Modality, and Relational Time," *Philosophical Studies* 70: 251–77.

Jan Cover's article presents an interesting criticism of Smith's article "The New Theory of Reference Entails Absolute Time and Space" (listed below) and its argument that sentences such as "John is asleep now, but might not have been" may show the substantival theory of time to be true.

*le Poidevin, Robin (1990) "Relationalism and Temporal Typology: Physics or Metaphysics?," *The Philosophical Quarterly* 40: 419–32.
This article introduces some new arguments into the debate about the connection of the relational theory of time with physics and metaphysics.
Newton-Smith, W. H. (1981) *The Structure of Time*, London: Routledge.
A good introduction to the philosophy of time that includes a lengthy discussion of the relational and substantival theories.
*Rynasiewicz, Robert (1992) "The New Theory of Reference Does Not Entail Absolute Time and Space," *Philosophy of Science* 59: 508–9.
A brief but ingenious criticism of Smith's argument about "now" and substantivalism in Smith's article, "The New Theory of Reference Entails Absolute Time and Space" (listed below).
Shoemaker, Sydney (1969) "Time without Change," *The Journal of Philosophy* 66: 363–81.
The classic article that introduced the argument for substantival time based on the possibility that the universe can undergo a freeze, where change ceases but time still passes.
*Smith, Quentin (1991) "The New Theory of Reference Entails Absolute Time and Space," *Philosophy of Science* 58: 411–16.
This article lays out in detail the argument about the connection between sentences such as "John is asleep now, but might not have been" and the substantival theory of time.
*— (1993) *Language and Time*, New York: Oxford University Press.
This book includes in part an argument for the substantival theory of time that is based on the connection between language and reality.
*Swinburne, Richard (1981) *Space and Time* (2nd edn), New York: St Martin's Press.
Swinburne develops Shoemaker's argument for substantival time in Shoemaker's article, "Time without Change" (listed above).
Van Fraassen, Bas (1970) *An Introduction to the Philosophy of Space and Time*, New York: Columbia University Press.
This book presents a carefully written defense of the relational theory of time.

Dialogue 4

Eternity

SOPHIA: I believe there is no such thing as an eternal being, God; there is only time, change and the universe.

IVAN: I think it is rationally acceptable to believe that God exists and that time and the universe are created by this eternal being. Admittedly, I have no proof that God exists, but I believe the theory of a creator of the universe can be explained in rational terms. A case in point is the concept of eternity. I believe this is a rationally coherent concept and that the theory of an eternal creator of the universe can be accepted by a rational person. I will acknowledge that it is not easy to arrive at a fully satisfactory definition of eternity, but I believe that if we discuss the issue at some length we will be able to develop a satisfactory definition.

SOPHIA: I have my doubts, but I am willing to discuss the issue with you.

IVAN: Let us begin with the traditional description of eternity offered by medieval philosophers, such as Boethius, Augustine and Aquinas. They expressed their conception of eternity with the famous sentence: "Eternity is the possession all at once of unlimited life." God is eternal in that he (or she or it) possesses all at once a life that has no beginning and no end.

SOPHIA: This description seems obscure at best, self-contradictory at worst. If God possesses his life all at once, this means he possesses all of it simultaneously. But this means that no one part of his life is later than any other part of his life. And that suggests his life is confined to a single instant. But I can't imagine how a person could live a whole life in a single instant, which has no duration at all.

IVAN: Many theists, both medieval theologians and present-day philosophers of religion, do not understand God's eternity in terms of God existing at a single instant. They say that God does not exist instantaneously but permanently. They adopt what may be called the *non-temporal duration* definition of eternity. Eternity is a *non-temporal duration* that has no limits, no beginning and end. To say this duration is not a *temporal* duration is to say that the duration has no parts that succeed one another. There is no part of the duration that is later than any other part. To say that it has duration means that it does not last for one instant only, but is *simultaneous with all instants*. Every instant of time in the created universe is present for God. It is as if God is on the top of a mountain and sees all of creation spread out below him; he sees all of it at once.

SOPHIA: I already see a difficulty with this theory. Suppose God is simultaneous with 1993 and also simultaneous with 1950. This implies that 1993 is simultaneous with 1950, which is an obvious contradiction (since of course 1993 is later than 1950).

IVAN: Could you explain this difficulty more fully?

SOPHIA: The difficulty is this. If two things X and Y are both simultaneous with a third thing, then X and Y are simultaneous with each other. In terms of your theory, how could God be simultaneous with both 1950 and 1993 if one of these years is earlier than the other? The only possibility is that God *once was* simultaneous with 1950, when 1950 was present, and *is now* simultaneous with 1993, which is now present. But this implies that God lasts through time and thus is not eternal.

PHIL: I see a second difficulty. Ivan said God endures without beginning and end, but that there are no parts – no successive instants – making up God's duration. This seems to be a contradiction. To endure means to last through time. If you last through time, then one part of your life will be later than another part. So you will occupy successive instants!

IVAN: Your objections to this particular definition of eternity are telling ones. They suggest that the theologian needs a richer concept of eternity. There is a second definition of eternity that I will call the *tenseless duration* definition of eternity. The basis of the second definition of eternity is the supposition that *the tensed theory of time* is true. [The tensed theory of time is discussed in detail in Dialogue 6 and pp. 181–4 of the Appendix.] The tensed theory of time implies that whatever is in time is either future, present or past. The opposite theory, *the tenseless theory of time*, implies that some events are earlier or later than other events, but that events lack the properties of being future, present or past. According to the tenseless theory, all events exist equally, regardless of whether they are earlier than today or later than today. Now let us suppose that the tensed theory of time is true. Given this, if something exists but is neither future, present nor past, then it exists outside of time.

Suppose that God's life has earlier and later stages in it, but that none of these stages is future, present or past. It follows that God exists outside of time.

ALICE: But if God's life has earlier and later stages in it, then God's life must occur in time. How can something exist timelessly if some of its parts are earlier than some of its other parts?

IVAN: Remember I said that on our second definition of eternity, we are assuming that the tensed theory of time is true, i.e., that time consists of a future, present and past. If God's life has no future, present or past stages, then this life is not in time. I will grant you that in a certain sense we may say that "God exists in tenseless time" and the universe exists in tensed time. However, since we are assuming that the tensed theory of time is true, then God's life in "tenseless time" is not really a life in time. Rather, it is outside of time. It is an eternal life.

ALICE: Let me summarize your second definition of eternity, the *tenseless duration* definition. Eternity involves (1) existing outside of time and (2) having a life that has earlier and later parts. How does this second definition relate to the medieval phrase you originally introduced, that eternity is "the possession all at once of unlimited life."

IVAN: Human beings exist in time and so we possess only one part of our life at a time. If one part of our life is present, the other parts are either passed away or still to come. But God possesses his life "all at once." This means that no part of his life is passed away or is future and not yet existent. All the parts of his life exist equally. The successive parts of his life are analogous to the successive states of the universe as they are described by the tenseless theory of time; they all exist equally, even though they are not simultaneous.

SOPHIA: There is a problem with this second definition of eternity. Consider the part of God's eternal life which consists of God's act of creating the year 300AD. Let us call this part of God's life the part P_3. P_3 is earlier than the part P_4 of God's life, which is God's act of creating the year 400AD. P_3 is simultaneous with 300AD. And P_4 is simultaneous with 400AD. But both 300AD and 400AD are *past* years, considering that the present year is 1993AD. This implies that the parts of God's life that are simultaneous with 300AD and 400AD, the parts P_3 and P_4, are also past. But if some parts of God's life have the property of being past, it follows that God exists in time.

PHIL: Why does the fact that P_3 and P_4 are simultaneous with past years imply that P_3 and P_4 are past?

SOPHIA: Because if something X is past and something else Y is simultaneous with X, it follows that Y is also past. If the part of God's life P_3 is simultaneous with the past year 300AD, then P_3 is also past. It follows that P_3 possesses the temporal property of pastness. God exists in time, and thus is not eternal. So the second definition of eternity as tenseless duration fails.

IVAN: Your objections are good ones. Let me move on to a third possible way to define eternity. I shall call this the *present instant* definition of eternity. Let us suppose that God's eternity does not involve duration at all. God exists only in *one instant*. This instant has no successive parts and so God lives his life "all at once." Furthermore, this instant is unlimited.

ALICE: How could a single instant of life be unlimited? Furthermore, suppose there is a tiny bug that is like a firefly and that exists for an instant only. Would not your third definition of eternity imply the absurd result that this bug is eternal?

IVAN: If God's life occupies one unlimited instant, then there are no instants before or after his life. This is what it means to say that the instant is "unlimited." By contrast, there are instants before and after the bug's life, so that prevents the bug's life from being eternal.

SOPHIA: This third definition of eternity seems applicable if time does not exist. But time does exist. So how could God occupy a single instant, with no instants before or after it, if time existed?

IVAN: I fail to see the problem.

SOPHIA: The problem is this. If time exists, then there are an infinite series of instants. If God exists at just one of these instants, then there are instants before and after the instant at which God exists, so he would not be eternal. God would have a fleeting existence, appearing and vanishing in a single instant of time.

PHIL: There is another problem with this third definition of eternity. God is supposed to be continually sustaining the universe in existence by his will. How could he do this if he existed for one instant only? At best, God can only sustain in existence the instant at which he existed.

IVAN: Both of you are making the same mistake. You are assuming that the instant at which God exists is a part of time. It is not one of the infinitely many instants that make up time. Rather, it exists outside of time. God exists in what the medieval theologians called the *nunc stans*, which means "the standing now." Beings who exist in time exist in the "moving now;" first the instant is future, then it is present, and finally it is past. Nowness or presentness keeps moving from one instant to later instants, casting all existents from the light of the present into the darkness of the past. But God remains in the eternal light of presentness. The now in which God lives remains standing; it has presentness and does not lose presentness. God exists in a permanently present instant.

SOPHIA: It seems to me that you are confusing several concepts together. If God's standing now "remains present," that implies he lasts for ever in time. Something cannot *remain* present unless it remains present through successive instants. "Remaining" implies lasting through time. But then God is in time and is not eternal! On the other hand, if God does not exist in a now that stands still but moves, then God also exists in time. So your third definition of eternity faces a dilemma. Either God is present and remains present or God is present and then passes away. Either way, God exists in time.

IVAN: Your objection again seems to be a good one. Let us try a fourth definition of eternity. Let us call it the *tenseless instant* definition. God exists at only one instant. But this instant is neither a standing now nor a moving now. This fourth definition is different than the second and third definitions in that it implies that *the tenseless theory of time* is true. The tenseless theory of time implies that each event, be it located in 1910, 1993 or 1999, exists equally as much as any other event. There is no distinction between the future, present and past in reality. Rather, all events exist equally, regardless of whether they are earlier or later than today or occur today.

Now consider the whole time-series, the whole composed of all instants. God exists outside this whole, but is conscious of each year that exists in the whole. Since God exists outside the whole, he does not exist at any temporal instant; he does not exist at any instant that stands in relations of earlier or later to other instants. Rather, he exists in a timeless instant. This instant is not earlier, later or simultaneous with anything; it is also not present, past or future. Since it is not a *present* instant it is not a "standing now." It is a *tenseless instant*.

ALICE: How is God related to the years that make up the history of the universe?

IVAN: Each year is present to him equally.

ALICE: You say "Each year is present to him equally." What do you mean by "present to"? Do you mean that God is timelessly present, and each year is seen by God to be temporally present?

IVAN: No. For two reasons. First, I am adopting the tenseless theory of time, according to which nothing has any properties of presentness, pastness or futurity. Events merely stand in relations of earlier, later and simultaneity. Second, if each year is seen by God to be temporally present, then we would have a contradiction. If God sees 1993 to be present and 1950 to be present, he would be mistaken. If 1993 possesses the property of presentness, and is seen to possess it, then 1950 must possess the property of no-longer-being-present, i.e., the property of pastness, and must be recognized to be past.

ALICE: I understand, but you still haven't explained the positive sense you attach to "Each year is present to God."

IVAN: I mean that each year is present to God in the sense that that year exists equally as much as any other year and that each year is apprehended equally by God.

ALICE: That makes sense to me.

SOPHIA: Let us not be too quick to give our assent. Ivan, would you summarize for us this fourth definition of eternity?

IVAN: God is eternal in the sense that (1) he exists outside of time in a single instant that has no temporal relations to any instant in time, (2) the tenseless theory of time is true, so that all years exist equally and (3) he is equally conscious of each year that exists.

SOPHIA: This theory of eternity seems to be the least problematic that you have presented so far. But I am not entirely satisfied with it. Part of your definition of eternity is that God exists outside of time. But God is also supposed to create each state of the universe. For example, God creates the state of the universe that exists in 1992 and the state that exists in 1993 and every other state. How can God do this if God exists outside of time?

IVAN: I fail to see any problem here.

SOPHIA: God exists outside of time, so he can't change. But God first creates the year 1992 and then, later, creates the year 1993. This implies that God changes. He first is in one state, *creating 1992* and then changes to another state, *creating 1993*. How can an unchanging being change?

IVAN: God does not need to change in order to create. God's many creative acts are not in time. They stand timelessly in relation to the successive years.

SOPHIA: But if 1992 is earlier than 1993, then God created 1992 before God created 1993!

IVAN: No. The premise *1992 is earlier than 1993* is true. But it does not follow that God created 1992 before he created 1993. Rather, God timelessly created the state of affairs, *1992 being earlier than 1993*. Indeed, God created timelessly the entire succession of years that makes up the universe. God timelessly creates the years 1992, 1993, etc., but also timelessly creates the temporal relations between the years. That is, God timelessly creates 1992 and 1993 and also timelessly creates 1992's being earlier than 1993.

SOPHIA: Let me try to understand this new theory of eternity more fully. It is now 1993. Does God exist simultaneously with 1993?

IVAN: If God did, then God would be existing in time. Since God does not exist in time, God does not exist simultaneously with 1993. But it is true to say in 1993 that God exists timelessly. This is different than saying that God exists in 1993. God does not exist in any year, but in each year it is true to say that God exists timelessly. Our true utterances about God exist in time, but what these utterances are about, God's existence, does not occur in time.

SOPHIA: This is a subtle theory, but it is beginning to make sense. However, I have a further difficulty. I will assume for the sake of argument your assumption that God exists.

Suppose, then, that at noon, I am worshipping God. At noon, God has the relation to me of *being worshipped by me*. But at 1 o'clock, I am thinking of something else and God no longer possesses this relation to me. Now if God does not exist in time, then how could he stand in this relation to me at one time and not stand in this relation to me at a later time?

PHIL: The fact that people worship God also shows that God undergoes changes. God changes from the state of *being worshipped by me* to the different state of *not being worshipped by me*. If God undergoes changes, then he must be in time. If God is first in one state, and at a later time is in a different state, then God is in time.

IVAN: But these are not *real* changes that God undergoes. They are real changes in you. Your state of mind changes from one of worship to a state of something else, say reading a poetry book. But changes in your state of mind do not produce any change in God. God remains unchanged.

SOPHIA: I think Phil has a point about God undergoing a change. Admittedly, it is a different kind of change than the one I undergo when I change from worshipping God to reading a poetry book, but it is a change nonetheless. I undergo an internal change, a change in my feelings, beliefs and ideas. But God changes externally, in the relations in which he stands to his creatures. At one time, he stands to me in the relation of being worshipped by me, and at a later time he does not stand in this relation to me.

IVAN: Suppose we say God "changes" in this external sense. This kind of change does not require God to exist in time. God's *mental states* must change in order for God to exist in time. If God changes from feeling pity to feeling contempt for the human race, then God undergoes an internal change and would exist in time. First he would feel pity and at a later time, contempt. But God does not change in this way. So he does not exist in time.

SOPHIA: How can you say that only *some kinds* of change require temporal existence? Surely, any kind of change requires temporal existence! If God is in one state at one time (be it internal or external) and a different state at a later time, then God has lasted through two times, one time at which he was in the first state and a later time at which he was in the second state. The fact that *being worshipped* and *not being worshipped* are external states of God makes no difference here.

IVAN: I think you are stretching the concept of time too wide if you want to say that God exists in time even if he undergoes no internal change. But even if you are right about time and external change, that proves only that there exists no eternal being, no deity that exists outside of time. But it does not prove that God does not exist. For God could exist in time. If God exists, God exists at all times. The common belief that God is "permanent" could be analyzed as meaning that he exists at every time, rather than that he is eternal.

SOPHIA: I have doubts that any notion of God (be it temporal or eternal) is rationally coherent, but I won't pursue the argument any further. I am satisfied with ending the discussion with the conclusion that the concept of an eternal being is rationally incoherent. As far as I am concerned, it is rational to believe that there is time, change and the universe, and nothing besides.

GLOSSARY OF TERMS

Eternity There are several definitions of eternity, but they share in common that eternity is God's timeless mode of existence.

God God is traditionally defined as a disembodied mind that is omniscient (all-knowing), omnipotent (all-powerful), perfectly good and the creator of the universe.

The "non-temporal duration" definition of eternity God endures simultaneously with all instants of time, rather than exists instantaneously, but his duration does not consist of stages that succeed one another.

The "present instant" definition of eternity God does not endure but exists only at one permanently present instant. This instant is not a part of time and is unlimited in the sense that there are no instants before and after it.

The "tenseless duration" definition of eternity God's duration possesses successive parts, but the parts of his duration do not possess properties of being future, present or past. This definition requires the tensed theory to be true, i.e., that events in time possess properties of being future, present or past.

The "tenseless instant" definition of eternity God exists at one instant that is not a part of time. This instant is not present and it is not simultaneous with, earlier than or later than any other instant. The tenseless theory of time is true, so all successive events exist equally and are equally apprehended by God.

STUDY QUESTIONS

1 What arguments can be brought against the definition of eternity as a non-temporal duration?
2 What arguments can be brought against the definition of eternity as a tenseless duration?
3 What arguments can be brought against the definition of eternity as a present instant ("standing now")?
4 What arguments can be brought against the definition of eternity as a tenseless instant?
5 Do you agree with Sophia that something exists in time if it undergoes external changes but does not undergo any internal changes?

FURTHER READING

*Leftow, Brian (1991) *Time and Eternity*, Ithaca, NY: Cornell University Press.

This book is the contemporary classic on eternity and will remain central to future discussions of eternity.

Lewis, Delmas (1984) "Eternity Again," *International Journal for the Philosophy of Religion* 15: 73–9.

Lewis argues that if God is observable in some sense by us, then God must exist in time.

— (1988) "Eternity, Time, and Tenselessness," *Faith and Philosophy* 5: 72–86.

Lewis offers an interesting discussion of the connection between eternity and the tenseless theory of time.

Padgett, Alan (1989) "God and Time: Toward a New Doctrine of Divine Timeless Eternity," *Religious Studies* 25: 209–15.

This article presents an interesting theory of divine eternity and its relation to Einstein's Theory of Relativity.

— (1993) *God, Eternity and the Nature of Time*, New York: St Martin's Press.

Padgett develops a novel theory of eternity and time.

*Smith, Quentin (1989) "A New Typology of Temporal and Atemporal Permanence", *Noûs* 23: 307–30.

This article includes a criticism of Stump and Kretzmann's article, "Eternity" (listed below) and offers a new definition of eternity in its place.

Stump, Eleonore and Kretzmann, Norman (1981) "Eternity," *Journal of Philosophy* 79: 429–58.

A much-discussed article that attempts to revive the traditional definition of eternity offered by Boethius and Aquinas.

Wolterstorff, Nicholas "God Everlasting," in S. Cahn and D. Shatz (eds) *Contemporary Philosophy of Religion*, New York: Oxford University Press.

Wolterstorff presents a plausible argument that God exists in time.

Part II

Time and identity

L. Nathan Oaklander

Dialogue 5

The problem of change

A college campus in New England.

PHIL: I find philosophy fascinating, but terribly puzzling, for it raises perplexing questions about what initially seem to be the most obvious features of our experience. I am reminded of the Irish philosopher Bishop Berkeley who claimed that philosophers raise the dust and then complain they cannot see.

ALICE: What features of our experience do you have in mind?

PHIL: There are two issues that are especially fundamental and give rise to numerous questions; time and change. Let us discuss change first. You and I know, or think we know, that things change. For example, I change, my friends change, my plants change, and so does just about everything that exists. For example, my tomato is now red, but it first was green. However, when I reflect upon the fact of change, I see a problem.

ALICE: Really, what is it?

PHIL: On the one hand, whenever there is change there is identity. It is *one and the same* tomato that is green and then red. In order for something to change, the thing must remain what it is, otherwise it would not be *that thing* that undergoes change. Take another example: I just painted the walls in my study. Before the change the walls were light blue, and after the change they were white. Still, it is the same wall that persisted through the alteration, that is, through the gain and loss of properties. On the other hand, whenever there is change, there is diversity: the thing must not remain what it is, but become different. Maybe I am just confused, but I seem to see a problem here. Change seems to require that the thing that changes be one thing or identical to itself, and two things or different from itself, and that strikes me as totally impossible. So, although I seem to perceive change, upon reflection I have my doubts concerning its reality.

ALICE: But how could you possibly doubt change? It is *obvious* that change occurs!

PHIL: I am not denying that change obviously seems to occur. I am just raising the question of whether what normally seems to us to be change may turn out not to be change at all, once we correctly analyze what is going on. Certainly reality seems to be changing, but isn't it a possibility that this appearance is deceptive and that reality is in some sense changeless?

ALICE: It is hard to take that possibility seriously. But I am interested in your idea. You said there was a problem about a thing being both the same and different if it changes. But I am not sure that I see any problem here. After all, isn't the tomato that is green the same as the tomato that is red?

PHIL: Yes.

ALICE: And isn't the property of green different from the property of red?

PHIL: Yes.

ALICE: Well then, we should say the tomato is the same as itself both before and after the change, and the properties that the tomato has are different before and after the change. It sounds paradoxical when you say "Change involves the tomato being what it is and being what it is not," but the air of paradox disappears once we distinguish the thing or substance, the tomato, from its properties. If a thing changes, then obviously it is not qualitatively identical with what it was because it has different properties or qualities from those it previously had. But even though it is qualitatively different it may still be numerically the same thing that undergoes the change.

PHIL: That seems simple enough. Change occurs when one and the same thing lacks a property that it formerly had.

ALICE: That's right.

PHIL: Wait a minute. I still see a problem since I don't think that you can have *one and the same thing* at two different times if the alleged single thing has different qualities or properties at those times.

ALICE: Really, why not?

PHIL: Let me pick up this old tomato here and put it between us. We are now both looking at it from different perspectives.

I see the back side and you see the front. Each of us may perceive the tomato to have some properties that the other does not perceive, for example, it may be partially rotten in the back, but not in the front. However, given we are looking at the *same tomato*, the tomato I perceive will have all the properties that you perceive it to have, and vice versa. Let me give another example before I draw the consequence and make my point.

If the chair I am looking at is the same as the one you are sitting on, then every property possessed by the chair I am looking at will also be possessed by the one you are sitting on. The general point is that if what appears to be two things are really identical (are really one thing), then every property of what appears to be the one thing will also be a property of what appears to be the other. Alternatively, we could say that if "two" things are really identical, then everything true of the one will also be true of the other.

ALICE: I accept that principle, but why does that cause you to doubt the reality of change?

PHIL: It seems to me that we are caught on the horns of a dilemma. Let's go back to the tomato. We are agreed that if the tomato changes, then the tomato which is green is *identical* with the tomato which is red. But we have just agreed that if the green tomato is identical with the red tomato then every property that the green tomato has will also be a property of the red tomato and vice versa.

ALICE: Yes, but what is the dilemma?

PHIL: Just this. If the green tomato has the same properties as the red tomato and vice versa, then the green tomato must be red, and the red tomato must be green, and that, if not an outright contradiction, is certainly impossible in some sense. On the other hand, if you deny that the green tomato has all the properties of the red tomato, then you have two different tomatoes and not one. In either case, change appears to be impossible.

ALICE: You must be making a mistake somewhere. Can't you see how excited I have become; can't you simply see I have changed?

PHIL: I see what you see, or rather I can see you are getting very agitated over this issue, which is good, but my reason tells me you have not really changed. I told you philosophy is puzzling. Maybe there really is change, but for the life of me I cannot see how it is possible. Can you see any mistake in my reasoning?

ALICE: As a matter of fact I can. You say, for example, I cannot change from being calm to being agitated and keep my *identity* because if I did I would have to possess incompatible properties, and that is contradictory or in some other way impossible.

PHIL: Yes, that's my point.

ALICE: But you have overlooked the obvious.

PHIL: Really, what obvious fact or truth have I overlooked?

ALICE: You have failed to distinguish between the essential properties of a thing and its accidental properties. An essential property of me is one that I cannot lose and still be the individual who I am. Thus, my being a human being is an essential property of me. I could not change that property and take on the property of being an insect (*pace* Kafka's *Metamorphosis*) and still retain my personal identity. Accidental properties, on the other hand, are those I can shed without ceasing to be who I am. Thus, my weighing 150 pounds is an accidental property of me since I may gain or lose a couple of pounds without losing my personal identity or ceasing to exist.

PHIL: Yes, there is such a distinction, but how does it help avoid the problem I am raising?

ALICE: What explains the identity of the tomato with incompatible properties is that the tomato which is green has the same essential quality of being a tomato as the tomato which is red. Similarly, I can change from being agitated to being calm and keep my identity since my essential qualities are the same whether I am calm or agitated even though my accidental qualities are different. It has *different* and incompatible properties, but it is *one* tomato that exists both before and after the change because it has the same essential form, namely, being a tomato.

PHIL: I don't think that solution works. The essence of this particular tomato is the same as the essence of that tomato over there. So, if the essence of the tomato is the basis for its identity then this tomato would be identical to the tomato over there, and we would still be stuck with a contradiction since this tomato is green and that one is red.

The problem with your way out can be viewed differently. It is agreed that if an object A persists through a change, then it must be identical to itself (= B) both before and after the change. You say that an object can be identical to itself – that A can be identical to B – so long as they have the same essential qualities, but I do not see why you limit the concept of identity to essential qualities. Admittedly, if A and B do not have the same essential qualities they are not the same object, but if A and B do not have the same accidental qualities they are not the same object either. Our concept of identity requires that if we are really dealing with one and the same object, so that A and B are really two names for the one object, then *everything* true of A will also be true of B.

ALICE: Could you explain that one more time?

PHIL: Yes. I agree that as we ordinarily think about things, an object can persist through an alteration of some of its properties, namely, those which are accidental, but cannot persist through a change in others, namely, those which are essential. But what I am saying is that we also ordinarily think that if what appears to be two objects is really one object, then *every* property that the one object has the other must have too. In this principle we do not distinguish between accidental and essential qualities. If the "two" objects are one then they must have the same accidental *and* the same essential qualities.

Go back to an example we considered a moment ago. This chair is what it is, and not a different object. Thus, any object that differs in any of its properties from this chair is not this chair. This chair is brown. Could a chair which is not brown be this chair? Could this chair be both brown and not brown? I think not. This principle, which is known as Leibniz's law, or the indiscernibility of identicals, is just as certain as the intuitive belief that objects persist through an alteration of accidental qualities. The philosophical problem of change arises, in part, because of the conflict between our intuitions concerning identity and change.

ALICE: OK, but now I see what is wrong with your argument: you failed to include *time* in your description of a thing's changing. I am calm and agitated at different times, the tomato is green and red at different times, and so on. The contradiction vanishes with the introduction of time.

PHIL: Maybe so, but I don't see exactly how the introduction of time helps. After all, how can a thing that changes its properties from one time to the next have all the same properties both before and after the change? I agree that change, if it is possible, involves time, but I don't see how the mere appeal to time makes change possible.

ALICE: As I said before, and this is obvious to me, when I change from being calm to being excited, I have the one property at one time and the other at another time. Now, and this is the key, the introduction of time enables me to have the same properties when I am calm as I do when I am agitated. For suppose that I am agitated at . . . Hmm, do you know what time it is?

PHIL: No, I don't wear a watch.

ALICE: It doesn't matter. Let's say that "I was calm at t_1" and "I am agitated now at t_2." It turns out that at t_1 and at t_2 I had the same properties. For at t_1 I possessed the properties of *being calm at t_1* and of *being agitated at t_2*, and at t_2 I have exactly the same properties. Thus, even though I changed my psychological state from one time to the other, it is also the case that every property that I possessed at one time is also a property that I possessed at another. Thus, the introduction of time renders intelligible the idea of *one thing changing* its properties.

PHIL: I must be dense, but I just don't get it. Let me go back to the beginning and quickly retrace our steps. Sometimes that helps when doing philosophy. Change requires an *alteration* in the properties of *one thing*. But if a thing's properties alter then either we have a contradiction, e.g., the tomato is green (all over) and red (all over); you are both calm and agitated, etc., or there is not *one* thing that has the incompatible properties. In either case we do not have change. That's my argument. You say that the introduction of time avoids the contradiction and preserves change since there is no incompatibility between your being calm at t_1, and being agitated at t_2. There is no threat to identity since all the properties of you at t_1 are also properties of you at t_2. Your solution reminds me of a remark of Kierkegaard when he spoke of the doctor who cured the disease by killing the patient. In your case you remove the problems by doing away with change.

ALICE: What do you mean?

PHIL: If at t_1 you have the properties of *being calm at t_1* and *being agitated at t_2*, and you have the very same properties at t_2, then how can you say that you have changed? On your view, individuals always have all the properties they ever will have.

Thus, once you introduce time, any paradox involved in the concept of change is avoided all right, but the cost is high; you do away with change altogether!

ALICE: I can tell that I am really getting in over my head, but let me take one more stab at it.

PHIL: OK, but before you do let me get out a book by a turn-of-the-century British philosopher by the name of John Ellis McTaggart. He is famous for his alleged proof of the unreality of time, and I bring him up because my argument is based on something I read by him. He is considering Bertrand Russell's account of change according to which a poker changes because there is a time when it is hot and a time when it is not hot. To this McTaggart replies:

> But this makes no change in the qualities of the poker. It is always a quality of that poker that it is one which is hot on that particular Monday. And it is always a quality of that poker that it is one which is not hot at any other time. Both these qualities are true of it at any time – the time when it is hot and the time when it is cold. And therefore it seems to be erroneous to say that there is any change in the poker.[1]

That sums up my argument, but you were saying that you had another way of understanding change. I am all ears. After all, I don't want to be a sceptic about change.

ALICE: Maybe time figures into the account of change in a different way. As time passes things change. So perhaps it is the passage of time that constitutes the crucial ingredient in change. Right now my talking belongs to reality and your talking does not. Your talking and my listening are events that have ceased to exist. But one moment from now, my talking will not belong to reality. Reality will consist of something different. Reality only consists of the present and the present is always changing. Granting the thesis that only what is present exists, perhaps we can understand change in terms of the coming into existence and ceasing to exist of the successive stages or events in the history of a thing.

PHIL: Hmm. "The coming into existence and ceasing to exist of the successive stages or events in the history of a thing"!? Now it's you who are moving too fast, but before we go any further into this labyrinth of time and change I have an idea. There are always some philosophy professors in the Philosophy

Department coffee room, so maybe we could drop by and see what they have to say about these issues.

NOTE

1 John McTaggart (1927) "Time," in *The Nature of Existence*, vol. II, Cambridge: Cambridge University Press, p. 15.

GLOSSARY OF TERMS

Accidental property A property is said to be accidental to a thing when that thing can exist without it.
Essential property A property is said to be essential to a thing when that thing cannot exist without it.
Indiscernibility of identicals The principle states that if "two" things are identical, then every property that is possessed by one is also possessed by the other, and everything true of the one is true of the other.

STUDY QUESTIONS

1 What is the problem of change?
2 Does the distinction between qualitative and numerical identity solve the problem of change?
3 Does the distinction between accidental and essential properties solve the problem of change?
4 Do you think that the introduction of time avoids the problem of change? Why or why not?
5 What is Alice's final account of change? What questions does it give rise to?

FURTHER READING

Broad, C. D. (1923) "The General Problem of Time and Change," in *Scientific Thought*, London: Routledge and Kegan Paul, reprinted Patterson, NJ: Littlefield, Adams and Co., 1959.
Argues that the problem of change requires the coming into existence of hitherto non-existent future events.
Haslinger, Sally (1989) "Persistence, Change, and Explanation," *Philosophical Studies* 56: 1–28.
A very clear exposition of how a conflict of intuitions gives rise to the problem of change.

Heller, Mark (1992) "Things Change," *Philosophy and Phenomenological Research* 3: 695–704.
States and defends a popular theory of changing things against objections.
*Hirsch, Eli (1982) *The Concept of Identity*, New York: Oxford University Press.
Concerned with identity through time, first with respect to ordinary bodies, then underlying matter and eventually persons.
*le Poidevin, Robin (1991) *Change, Cause, and Contradiction: A Defence of the Tenseless Theory of Time*, New York: St Martin's Press.
Argues that the problem of change can only be avoided if we adopt a particular (tenseless) theory of time (to be discussed in Dialogue 6).
*McTaggart, John (1927) "Time," in *The Nature of Existence*, Cambridge: Cambridge University Press, vol. II, bk V, ch. 33.
In this famous chapter, McTaggart argues that time and change are unreal.
*Merricks, Trenton (1994) "Endurance and Indiscernibility," *Journal of Philosophy* 91: 165–84.
Merricks gives a careful account of the apparent conflict between change, indiscernibility and endurance, and then offers his own solution.
Oaklander, L. Nathan (1987–8) "Persons and Responsibility: A Critique of Delmas Lewis," *Philosophy Research Archives* 13: 181–7.
Discusses some of the relations among time, change and identity.
Smith, Quentin (1993) "Change," in Jaegwon Kim and Ernest Sosa (eds) *A Companion to Metaphysics*, Cambridge, Mass.: Basil Blackwell.
Smith summarizes the standard definitions of change that are discussed in the philosophical literature.

Dialogue 6

The passage of time

Walking to the coffee room of the Philosophy Department in the English–Philosophy Building.

ALICE: Our discussion of time and change reminds me of a conversation I had with my Dad not too long ago. He was saying that when he was an undergraduate there was a lot of discussion about the NOW generation. It was claimed to be important to live for the NOW; make every present moment count; don't worry about the future, it will take care of itself and so on. If people really thought like that then they might live their lives differently. For example, they might be less concerned about the consequences of their actions. On the other hand, if someone thought the future is as real as the present, or if they denied the NOW is some special point in time, then they might look at their life differently. They might forego immediate pleasures or gratifications for the sake of future ones. Thus, the way we think about time and the NOW, e.g., whether or not *only* the present exists or is real, could influence how we think about our lives.

PHIL: All the more reason to search out the truth in these matters.

ALICE: I agree.

PHIL: Did you know that a couple of philosophy professors, Ivan and Sophia, have written extensively on topics in the philosophy of time?

ALICE: Yes, and I expect they will be in the coffee room. Let's go.

In the coffee room, where they meet Ivan and Sophia.

IVAN AND SOPHIA: Hello.

PHIL: Alice and I have been discussing the problem of change and have come to something of a log-jam. We were wondering if either or both of you could shed some light on the question of whether or not change is real?

SOPHIA: Good question. Although the fact that you raise it leads me to suppose that either you have been reading Parmenides, or you have been taking too many courses from Ivan.

IVAN: You're making fun of me, Sophia.

SOPHIA: Just a little, but even though you don't believe it, your views on time could lead someone to believe that nothing *really* changes.

IVAN: Maybe, but before we get into that let's hear what these budding philosophers have to say.

SOPHIA: I agree. So, Phil and Alice, tell us what perplexes you about the concept of change.

PHIL: This is my problem. I hope you do not mind if I use a personal example. It helps me to focus the issue.

SOPHIA: Not at all.

IVAN: Please do.

PHIL: As I was growing up my parents always wanted me to be a doctor, I mean a *medical* doctor, and at one point in my life I thought that I wanted to be a doctor too. However, after taking a few courses in chemistry and philosophy I decided, much to the chagrin of my family, that I wanted to study philosophy and not medicine. In other words, I seemed to have *changed* my mind about what I wanted to be. But then, after having taken a few more philosophy courses and having read some McTaggart, I thought to myself: my wanting to be a doctor at one time and my wanting to be a philosopher at another time are not properties that I acquire and shed. Thus, it was, and still is, not clear to me in what sense I have changed from then to now, if it always was, is, and will be true that I possess these properties.

ALICE: There may be a problem here, but I think it can be resolved if we think of time in terms of temporal becoming. Thus, we can say that Phil did change because he *formerly* wanted to be a doctor, and he *now* wants to be a philosopher. What I mean is this. When Phil wanted to be a doctor, his want was present, it existed, but his wish or desire to be a philosopher did not yet exist, it was in the future. *Now* his wanting to be a philosopher is present, it does exist, but his wanting to be a doctor is past;

it no longer exists. Thus, as I said yesterday, it is the coming into existence and the ceasing to exist of successive stages in the history of a thing that gives rise to change.

I believe there is something special about those events that are present. For out of all the events in the history of the universe only those which are present enjoy the privileged status of being for a moment Real, and Existing with a capital E. All others have either ceased to exist, and so possess the property of *pastness*, or are yet to be born, and so possess the property of *futurity*. Indeed, the essence of time and change is to be found in the passage of time, or the passage of events through time, from the future to the present and from the present into the past.

PHIL: My problem with Alice's explanation of change is that I do not understand the notion of time's passing. It is unclear to me how the notion of not yet existing future events moving toward the present, momentarily becoming present and then receding into the past as they cease to exist, really explains anything.

ALICE: This is just about where we stopped yesterday. Does any of this make sense to you, or are we totally confused?

SOPHIA: You are not totally confused, although it is easy to get that way when one begins to reflect on time and change. In fact, there was one twentieth-century British philosopher, C. D. Broad, who in his fifty years of reflecting on the problems of time adopted no less than five different views on the nature of time. No wonder he thought the problem of time "is the hardest knot in the whole of philosophy."[1]

IVAN: I agree. Time is a thorny issue. I like the way St Augustine put it. He asked, "What, then, is time?" and replied by saying something to the effect that, if no one asks me I know, but if someone asks me to explain it I can't.[2] Still, I think we can make some headway, and shed some light on your dispute concerning the problem of change if we begin with what we know, or at least with what strikes us as truisms about time and see where that leads us.

PHIL: Sounds great.

ALICE: What are some of the truisms you have in mind?

SOPHIA: There are two different ways in which we ordinarily conceive of time, and philosophical problems arise when we reflect on these two different ways since it may appear that they cannot both be true of reality.

IVAN: On the one hand, we think of time as involving events strung out along a series united to one another by the relations of *earlier*, *later* and *simultaneity*. For example, my talking is simultaneous with Sophia's listening, and my talking is later than my breakfast this morning, and earlier than my dinner this evening. When we think of time in this way, time is like a very long, perhaps infinitely long, string of beads on which all events are eternally fixed. The events in the temporal series are fixed in that they never change their position relative to each other. For example, since Plato was born before Aristotle was born, it will always be the case that Plato's birth is earlier than Aristotle's birth. And since Augustine was born after Aristotle, it will always be the case that Augustine's birth is later than Aristotle's. It has become customary to call the entire series of events spread out along the time-line from earlier to later, the "B-series." When viewed solely in terms of the B-series, time is thought of as *static* or unchanging for there is nothing about temporal relations between events that changes.

SOPHIA: There is another way in which we conceive of time; a way I think is even more fundamental. Time not only has a static aspect, it also has a *transitory* aspect. In addition to conceiving of time in terms of events standing in temporal relations, we also conceive of time and the events in time as *moving* or *passing* from the far future to the near future, from the near future to the present, and then from the present they recede into the more and more distant past. Thus, for example, my dinner is now future, but with the passage of time it will become present and will become past. When events are ordered in terms of the notions of past, present, or future they form what is called an "A-series." It should be noted, of course, that the A- and B-series are not really "two" different series of events, but the same series ordered in two different ways.

IVAN: When philosophers wonder about the nature of time, one question they ask is which of these two conceptions of time (or which of the two series), if either, is more fundamental? In other words, are temporal relations the ultimate realities in terms of which time and change are to be understood, or are temporal relations analyzable in terms of the application to the world of the more basic concept of temporal passage and the properties of *pastness*, *presentness* and *futurity*?

ALICE: Why do we have to choose between one or the other? Since we ordinarily believe that events in time are temporally related and undergo temporal passage, shouldn't we say both the B-series (temporal relations) and the A-series (monadic temporal properties) are equally fundamental; that they both exist?

SOPHIA: Good question. The answer to it is debatable. Some philosophers reject tense entirely, and believe that temporal properties are a myth. These philosophers, called "detensers," believe our ordinary ways of speaking, conceiving and experiencing time can be understood solely in terms of the B-series of temporal relations. For example, Bertrand Russell denied that *being present* was a property that events possessed, and claimed instead that an event is present only if it is perceived or simultaneous with an object of perception.

Others maintain B-relations are analyzable or definable in terms of the notions of past, present and future because the passage of time is required to make the earlier/later relation a temporal one. Thus, McTaggart claimed that while both the A-series and the B-series are essential to our ordinary thought and experience of time, the A-series and temporal becoming is more fundamental to the real metaphysical nature of time since temporal or B-relations are dependent upon temporal becoming (or A-properties).

ALICE: What do you mean B-relations are dependent upon becoming?

SOPHIA: Well, he claimed you could define "earlier than" by means of A-expressions like "past," "present" and "future," but that A-expressions could not be defined in terms of B-expressions like "earlier than." For that reason he could be interpreted as maintaining that events can stand in temporal relations to each other only if they have temporal properties.

Some philosophers who believe in the tensed theory of time even maintain that there are neither temporal relations, nor temporal properties or events, but only substances and present tensed facts like *John is now happy*. And A. N. Prior argues that time consists of tensed facts such as *It was the case that John is fat*, *John is happy* and *It will be the case that John is angry*. But I find his theory unsatisfactory, if only for the reason that he never explains the constituents of these tensed facts. For example, if there is no property of pastness then what is the meaning of "was" in "it was the case that"? Prior and his followers have not given a satisfactory explanation.[3]

IVAN: As you can see, Alice, there are a variety of views on the topic of time. There is, however, some reason to believe that if time is to be real then the static and transitory conceptions of time cannot both be fundamental.

ALICE: Could you explain the reasoning to me?

IVAN: I'll try. In so far as we conceive of time in its transitory aspect, as *becoming*, we are conceiving of it from a given temporal point of view. What is real, what exists, are those events that exist *now, at this present moment*. Past events did exist, but exist no longer and future events, even if they will exist, do not yet exist. Thus, according to our ordinary conception, temporal becoming is the continual coming into existence of what did not previously exist and the continual going out of existence of what presently does exist. Given this common-sense view of things, past and future events are considered not to exist in the sense in which present things are said to exist.

On the other hand, when we conceive of time in its static aspect, as involving unchanging temporal *relations* between events, we are viewing time from a point of view outside of time. From this God-like perspective all events are equally real, having the same ontological status, and in some sense "co-exist" in the network of temporal relations that constitute the history of the universe. Given this conception, there are no ontological differences between past, present and future events.

These two aspects of our ordinary conception of time can easily give rise to puzzlement. For it is difficult to understand how all events, including those earlier and later than the present, can in some sense "co-exist" if only the present exists. To put the same point differently, how can a present event E_2 be later than a past event E_1 and earlier than a future event E_3 when temporal relations, like all relations in general, relate existents to existents and neither the past event E_1 nor the future event E_3, but only the present event E_2 exists?

PHIL: Let me see if I am following you. There is a difference between a pearl necklace and a single pearl. When we view time in terms of temporal relations, events in time are strung out like pearls on a necklace with no distinction between them. When we view time in terms of temporal passage the only pearl that really exists is the one which is present, and that one is constantly being replaced. Since the whole necklace cannot exist if only a single pearl does, the whole B-series cannot exist if only one momentarily present section does, and vice versa.

IVAN: Yes, that's the idea. My own view is that temporal relations are the fundamental realities and temporal passage and change can be explained in terms of them. I agree that we speak and think of time in terms of temporal passage, but I maintain that the reality that underlies the language of time and change involves nothing more than temporal relations between events. I have never found the notion of time's passing as anything more than a metaphor and although an appealing one, I doubt it represents any special fact about reality.

SOPHIA: I disagree, viewing the matter in precisely the opposite way. For me, temporal passage is the ultimate temporal reality and temporal relations are to be understood in terms of temporal passage. As I see it, the reality that underlies our ordinary conception of the B-series is not like the string of pearls model suggests. All the terms in the history of the world do not exist all at once, so to speak. Rather, only a single section of the string exists, namely, the section that is present. The future parts of the string will become present, they will come into existence, but do not exist yet, and the past parts of the string are those which did exist, but have presently ceased to exist. All that really exists is what is NOW, and as time passes new events take on the property of *presentness* and other events shed it for the property of *pastness*. Indeed, it is just this coming to be present and ceasing to be present that is the foundation for time and change.

ALICE: So the issue between you and Ivan comes to this: you both agree that as we ordinarily speak and think, time involves a static and a transitory aspect. However, Ivan maintains that temporal relations constitute the essence of time and change, and that there are no transitory temporal properties. Sophia, on the other hand, maintains that the transitory aspect of time is fundamental in that there are non-relational temporal properties and temporal relations can exist only if their terms possess A-characteristics and change with respect to them.

IVAN: Exactly.

ALICE: I'm glad I too understand something, but what Phil and

I want to know is which view is correct? How would you even go about resolving a disagreement over such a basic issue? And while you are answering those questions, could you please explain the precise connection between these different theories of time and the question Phil posed to you a few minutes ago, namely, whether or not change is real?

IVAN: Let me take up the methodological question first. Several different methodologies have been employed. Until quite recently the dispute between the two theories of time has centered around temporal language and the *method of translatability*. Defenders of the tenseless theory of time (so-called because its proponents maintain that all events exist regardless of whether they are past, present or future) used to argue that since tensed discourse could be eliminated or translated without loss of meaning into tenseless discourse, an adequate account of the nature of time need not countenance any special kind of (irreducibly) tensed fact such as *I am hungry now*. In other words, proponents of the tenseless theory maintained that an analysis of ordinary language that eliminated tensed discourse supported an analysis of temporal reality that eliminated temporal passage. For example, suppose that time could be completely described by such sentences as "1993 is later than 1992." We would not need sentences such as "1993 is present" and "1992 is past." But since we often use tensed sentences such as "1993 is present," the proponents of the tenseless theory need to show that tensed sentences have the same meaning as tenseless sentences.

PHIL: Could you give an example of what such an analysis would look like?

IVAN: One such account is known as the token-reflexive analysis. According to it, "X is present" means or is translated by "X is simultaneous with this token" where "this token" refers to the sentence "X is present" just uttered, written down or thought. Similarly, "X is past (or future)" is translated by "X is earlier (or later) than the sentence token 'X is past (or future).'" Thus, on the token-reflexive analysis of the meaning of tensed discourse, whether an event is past, present or future depends solely on whether or not it is earlier, simultaneous with or later than the sentence that refers to it.

ALICE: Does that mean that no event would be present unless somebody said that it was present?

IVAN: Yes it does.

ALICE: I find that weird. It seems obvious to me that I could be agitated right now even if nobody said that I was.

SOPHIA: I agree. That is one reason why I cannot accept a token-reflexive or a psychological analysis.

PHIL: What do you mean by a "psychological" analysis?

SOPHIA: Bertrand Russell, and C. D. Broad in his earliest writings on time, adopted a psychological analysis of the tenses. They claimed that the notions of past, present and future are analyzable into the temporal relations of states of mind and their objects. Thus, for example, to say "Event E is present" means, on this analysis, that "Event E is an object of perception or it is simultaneous with an object of perception." Analogously, "Event E is past" means "It is earlier than a memory of it," and to say "Event E is future" means "Event E is later than someone's anticipation of it."[4]

ALICE: Isn't this analysis open to the same type of objection as the token-reflexive account?

SOPHIA: Yes, for it is perfectly consistent to say that "Event E is present although it is neither perceived nor simultaneous with any object of perception." And yet, if the psychological reduction were true, then that sentence would mean "Event E is perceived or simultaneous with an object of perception, and it is neither perceived nor simultaneous with any object of perception" which is a contradiction.

IVAN: I suppose that is one reason why detensers have offered other, less subjective, analyses of tensed discourse. Another such attempt is known as the "date-analysis." According to it, a sentence in which the word "now" or its equivalent is used can be translated through the use of a second sentence formed by replacing the "now" in the first sentence with any date expression used to refer to the time at which the first sentence was uttered, written or even thought. Thus, for example, to say "Alice is agitated now," uttered at twelve noon, May 2, 1991, means "Alice is agitated at twelve noon, May 2, 1991." Similarly, the sentence "Alice will be agitated," if uttered at twelve noon, May 2, 1991, means "Alice is agitated later than twelve noon, May 2, 1991."

SOPHIA: It is hard to believe that these tenseless translations can capture the meaning of tensed discourse either.

ALICE: Why?

SOPHIA: One line of reasoning goes like this. If two sentences have the same meaning then all the information conveyed or communicated by the one sentence should also be communicated by the other.

PHIL: That seems correct. For example, since "There is a bachelor in the corner" means (roughly) the same as "There is an unmarried male in the corner," either of these sentences could be used to convey the same information.

SOPHIA: The problem with date-analysis translations is that they do not convey the same information as their tensed counterparts. Suppose I am sitting in a dentist's chair, wanting to convey to him the information that my tooth hurts now. I will not be able to do that by telling him my tooth hurts at 3:00 p.m., since the dentist may not know that it is *now* 3:00 p.m. In order to convey the information of the original sentence ("My tooth hurts now") I would have to add the additional information that *now* is the time my tooth hurts, and that brings tense back into the meaning of the sentence. Thus, it would appear that tense is ineliminable from ordinary discourse and consequently, temporal passage is ineliminable from reality.

IVAN: Well, I did say that defenders of the tenseless view *used* to use the criterion of translatability as their methodological principle, but for reasons you suggest, and others, they seldom use it any more.

ALICE: What other reasons do detensers give for rejecting translatability as the mark of a correct analysis of time?

IVAN: Recent defenders have argued that tensed discourse is indeed necessary for timely action, but temporal passage is not since the reality underlying tensed sentences can be represented by a tenseless language that describes unchanging temporal relations between and among events. In other words, it is agreed that ordinary language and thought require tense, but it does not follow that the proper account of the nature of time requires the objective reality of temporal properties such as presentness and pastness. Indeed, defenders of the new tenseless theory of time have gone further and argued that tensed discourse cannot represent the ultimate nature of time because the metaphysics (as opposed to the language) of tense is contradictory. As far as ordinary language is concerned tense is unproblematic and indispensable, but if we want to represent the true nature of time, then a metaphysically more perspicuous language should be used and it should be tenseless.

ALICE: I don't understand. Why would anybody want to deny that events are really or intrinsically present in favour of the idea that events simply occur (tenselessly) at certain times? What is the absurdity involved in the reality of tense?

IVAN: To answer your question we need to turn to another method that philosophers have employed in the dispute between the tensed and tenseless theories of time, namely, the *dialectical method*. Reflection on the logical implications of each of the two views has led some to adopt one position rather than another.

PHIL: Does the dialectical method favor the tensed or the tenseless view?

IVAN: It all depends on who you ask. I cannot accept the tensed view because it raises serious dialectical difficulties, and I am sure that Sophia thinks the same about the tenseless view.

SOPHIA: Yes, I do.

PHIL: What dialectical difficulties do you find in taking temporal passage to be real, Ivan?

IVAN: There are many, but the most relevant to our discussion is that I do not think the tensed view really helps to explain how change is possible.

SOPHIA: Really, why not?

IVAN: Tell me, Sophia, do those events that are present change or are the same events always present?

SOPHIA: Obviously, as time passes different events become present. Indeed, if those events now present did not change, then nothing else could change. That is, if what exists now did not change, then we would be frozen in the present moment and change would be impossible. So, yes, events change and they do so constantly as time passes and new events come into existence.

IVAN: Then the problem of change applies all over again to events. If events change from being future (or not yet existing) to being present (or existing), to being past (or ceasing to exist), then one and the same event is what it is not and is not what it is, and the problem of change is to understand how that (contradiction) is possible.

SOPHIA: That's easy enough to answer. An event changes by having different temporal properties at different times. What I mean is this: if E occurs omnitensely (i.e., occurs, will occur, occurred) at t_1, and t_1 is present, then E is present. If E occurs at t_1 and t_1 is past, then E is past; and if E occurs at t_1 and t_1 is future, then E is future. This analysis removes any contradiction in event E having incompatible temporal characteristics and it preserves the reality of change.

IVAN: Actually, I am not sure that it does either of those things. Your analysis of "Event E is present" involves the tensed fact that *t_1 is present*, but that fact must itself *change* from obtaining to ceasing to obtain; it first acquires and then loses the property of obtaining. Furthermore, on your view the propositions expressed by, say, "Event E is happening now" or "It is now 1993," change their truth value with the passage of time, first becoming true and then later becoming false. It appears, then, that we are still left with the question of how tensed facts and tensed propositions can retain their identity through a change of properties. Thus, as I see it, to explain change in terms of temporal becoming is circular because temporal becoming or the passage of events, facts or propositions through time is a kind of change.

ALICE: Wait a minute. You are going too fast. Will you run through the argument one more time?

IVAN: To me, Sophia's view of time and change is going around in circles, and it never really explains how change is real or even possible. For to explain a *thing's changing* its properties she introduces time in the form of temporal becoming. Then she explains temporal becoming or passage in terms of *events changing*, and to complete the circle she explains how events can change by introducing temporal becoming once again in the form of *tensed facts changing* their "obtaining status," or *propositions changing* their truth value.

I should add that not only is the tensed theory of change circular, it is also contradictory. On Sophia's analysis, the explication of the sense in which *events* change their temporal properties relies on *times* changing their temporal properties. That is, t_1 (and presumably each other time) is past, present and future. Thus, the contradiction we found in an event undergoing temporal becoming recurs again at the level of times at which events don and doff their temporal properties. For these reasons, I do not believe the tensed view provides an illuminating and consistent theory of time and change.

SOPHIA: You raise many interesting issues and I don't think we should get into them now, but I must point out one basic flaw in your reasoning. You say that my account just transfers the contradiction from one set of terms, events, to another set of terms, moments.

IVAN: Yes, that's right.

SOPHIA: Well, I disagree. The mistake in your reasoning is to suppose that it is ever true that "t_1 (or event E) is past, present and future." Clearly, that is false. What is true is that "t_1 is past, *was* present and *was* future," or "t_1 is present, *was* future and *will be* past," or "t_1 is future and *will be* present and past." There is no contradiction in each of the propositions expressed by those statements being true.

The point can be put slightly differently with the emphasis on the idea that only present events and times really exist. Thus, since 1993 is present, for example, it is true to say "1993 is present." Of course, it is also true to say, "It will be true to say 'It was true to say that "1993 is present."'" And it is also true to say that "It was true to say that 'It will be true to say that "1993 is present."'" What we should not say is that "1993 is past, present and future." On my view that is never true. Once that is realized, no contradiction in temporal becoming will result.

IVAN: Like just about everything else in philosophy what you say is debatable. For whether your way out of the difficulties I am raising is anything more than a verbal maneuver depends upon the ontological interpretation that is given for each of the propositions you say is true. What, for example, is the reality that underlies the truth of a claim like: *What is present was future, and will be past*? That is one question I would like clarified before I would consider your view to be plausible. Since, however, these issues are rather complex maybe we should not discuss them further right now.

SOPHIA: Agreed. But tell me, if you reject the argument for the reality of change based on temporal becoming, do you believe that nothing really changes?

IVAN: Not exactly. If you think of reality as signifying everything that exists, a *whole* of which things and events are in some sense "parts," then my view is that reality is changeless. The whole string (the entire B-series) does not change. Nevertheless, I do not deny the reality of change, for I believe that though *reality itself does not change* – it is what it is and never another thing – *things that exist in reality do change*. In other words, change does not apply to the B-series taken as a whole, but the B-series contains things that do change.

PHIL: In what sense do things change, according to you?

IVAN: Unlike you Phil, I think events like *Alice's being calm* or *Alice's being agitated*, and facts like, *Alice's being calm at* t_1 and *Alice's being agitated at* t_2, are sufficient to account for change. Admittedly, neither the *events* nor the *facts* in question change, but just because they don't change it does not follow that no *thing* changes. To say a thing changes means nothing more and nothing less than to say there exists a succession of events (of one thing) with qualitative similarities and differences.

SOPHIA: What makes a succession of events a succession of *one thing* should not be glossed over too quickly in a discussion of change, but since the focus at the moment is on time and not identity I will let that issue pass. Even if we set the identity issue aside, I still don't see how a thing's having a property at one point in a series and a different property at another point can constitute change. After all, different colored objects at different spatial locations do not give us change, so we need some further explanation of why qualitatively different events in a series do give us change. On my view, becoming or temporal passage is the foundation of events standing in temporal relations. What is the foundation for time on your view?

IVAN: First, I would say that given the dialectical difficulties I raised concerning temporal passage, I don't see how passage can be the foundation of a series being temporal, i.e., being ordered by the relation of before/after. As to your question, "What distinguishes a spatial series of qualitatively different parts from a temporal series of qualitatively different events?", ultimately I think the answer is to be found by consulting your experience. Listen to me snap my fingers. You have just had the immediate experience of one event's following after another. You might say that we have this experience of succession because of the passage of events from present to past, but I disagree. As I see it, the experience of earlier and later upon which the experience of time and change is built is primitive and intrinsically different from our experience of one colored patch being to the left of another differently colored patch. The phenomenology of the situation acquaints us with simple, unanalyzable temporal relations and from it our concept of a temporal series is derived.

You mentioned C. D. Broad. Did you know that in his earliest musing on time he was a detenser? My views on our experience of time is reflected in something he once said. Here, listen to this:

> Temporal characteristics are among the most fundamental in the objects of our experience, and therefore cannot be defined. We must start by admitting that we can in certain cases judge that one experienced event is later than another, in the same immediate way as we can judge that one seen object is to the right of another . . . On these relations of before and after which we immediately recognize all further knowledge of time is built.

This is the view of our experience of time I wish to adopt.

SOPHIA: Your appeal to how we experience the world, or as you also put it, the phenomenology of the situation, can hardly strengthen your view of time. Indeed, it undermines it! For if we turn to a third methodology employed in the debate, the *phenomenological method*, the weakness of the tenseless view becomes most evident.

ALICE: What is the phenomenological method?

SOPHIA: According to this method, any adequate account of the world must accord well with how we experience the world, and with our commonly accepted attitudes and perceptions of the world. Now, it seems to me that there are a host of attitudes and perceptions about the world that can hardly be explained on the tenseless theory of time, but can easily be explained on the tensed theory.

IVAN: What attitudes do you have in mind?

SOPHIA: For one, the tenseless view cannot adequately explain some of our different attitudes toward the past and the future. For example, a very painful experience that is known to have happened to us in the past, say, a painful visit to the dentist, is, when contemplated, thought of with *relief*. As we sometimes say, "Thank goodness that's over!" On the other hand, when we know that an equally painful experience is something that will occur in the future, our attitude is one of dread and anxiety. Similarly, we often feel nostalgic over pleasant events that have happened and joy over events that we expect to happen. In both sorts of cases, there seems to be a kind of conceptual connection between our different attitudes and the nature of time. Events contemplated with nostalgia are not just those that are past, but those that are receding or moving away from us. Those that are contemplated with joy are those that are understood to be coming towards us and about to overtake us; about to be in the NOW, in the stream of our lived experience. These are undisputable phenomena that can be explained on the tensed theory of time because on this view events really are moving toward us. On the tenseless view, nothing really moves through time and so such phenomenological facts seem inexplicable.

PHIL: Why can't the tenseless view explain the phenomena you mention?

SOPHIA: The problem I have is this: if all events in the temporal series are equally real, and there are no basic distinctions between past, present and future events, then dread and relief are never appropriate attitudes to take toward a painful tooth extraction. For one thing, if no events are really past or future, how can I take any special attitude toward past or future events? For another, consider the following reasoning. Suppose I am having a tooth extraction on Tuesday, then on the tenseless view relief would be appropriate on Wednesday because Wednesday is later than Tuesday. But the fact that *Wednesday is later than Tuesday* is an unchanging fact that *always obtains*, i.e., before, after and during the extraction. Thus, if that fact explains the relief, then for the painful experience that is taking place on Tuesday it is as sensible to feel relief on Monday or Tuesday as on Wednesday, and that is absurd.

IVAN: I agree that we do have an experience of time's flow and our different attitudes toward one and the same event reflect this experience. Nevertheless, I think that the tenseless theory can render intelligible both our experience and the attitudes. Consider the dreaded visit to the dentist. Now, at time t_1 I anticipate going to the dentist, and that anticipation is earlier than the event in question. At a later time, t_2 I have the perception of being in the dentist's office, sitting in the chair, feeling him pick at my teeth, etc. This perception is simultaneous with the formerly dreaded event. Finally, at a later time I have the memory of the entire dreaded experience, and because my memory is later than the experience I feel relieved.

It is just this succession of different psychological attitudes toward the same event (first anticipation, then perception, then memory) together with the succession of the different experiences themselves that gives rise to the impression of time's flow, and it is that impression that provides the basis for our different attitudes toward the future and the past.

SOPHIA: I still don't see how these different psychological attitudes can explain our feeling of relief unless events *really* cease to exist. Maybe I can get at what troubles me in another way. Tell me, can you know what it is like for a headache of yours to cease to exist?

IVAN: Yes, I can.

SOPHIA: Do you also maintain that all events co-exist in the network of temporal relations, i.e., in the B-series?

IVAN: Yes.

SOPHIA: Well then, my argument is that if all events co-exist (or exist tenselessly), then you cannot possibly know what it is like for a headache of yours to *cease to exist*. Since you do know this, it follows that your view of time is mistaken.

IVAN: I agree that if the tenseless theory of time implied that I cannot know what it is for a headache of mine to be over then there would be a problem with the tenseless theory, but it does not have such an implication. On the tenseless theory, the tenseless facts are: (1) I am (tenselessly) conscious of having a headache at t_1, (2) I am (tenselessly) conscious of taking an aspirin (and having a headache) at t_2 and (3) I am (tenselessly) conscious of feeling fine (and not having a headache) at t_3. The succession of these different states of consciousness is the basis for my knowing that a headache of mine has ceased to exist. Your argument is based on the erroneous idea that since the tenseless view maintains that all events exist, there can be no ceasing to exist. But on the tenseless view an event can cease to exist. To say that an event ceases to exist means that there is a time after the existence of all of its temporal slices (or after its temporal location), or that it occupies a moment before or after my statement about it.

It seems to me that the issue between us centers around the correct interpretation of "ceases to exist." On your view only the present exists; the past no longer exists and the future does not yet exist. Thus, when a headache ceases to exist it "really" ceases to exist. Of course, *assuming* that interpretation of "ceases to exist," a detenser, such as myself, cannot know that my headache has ceased to exist since nothing ever does *really* cease to exist. But such an interpretation assumes what needs to be proved. In other words, if we understand "ceases to exist" as Sophia would have it, then it is far from obvious (to me at any rate) that we do know what it is like for a headache to cease to exist.

On the other hand, if we assume the tenseless interpretation of "ceases to exist" then we *can* know that a headache has ceased to exist. According to the detenser, a thing ceases to exist if and only if there is a time after the existence of all of its temporal slices (or after its temporal location). In short, it does not seem to me that I am aware that something *really* ceases to exist (in the tensed sense, whatever exactly that may be) and so there is no item of knowledge that the tenseless view cannot explain.

SOPHIA: Obviously we could go on with this debate, but I am getting hungry. Let's say we go get something to eat and continue our discussion after lunch.

PHIL: Sounds like a good idea to me.

IVAN AND ALICE: OK, where shall we go to eat?

NOTES

1 C. D. Broad (1959) *Scientific Thought*, New Jersey: Littlefield and Adams, p. 84.

2 St Angustine (1948) *Confessions*, trans. E. B. Pusey, Chicago: Encyclopedia Britannica, Book XI, section XI.

3 Bertrand Russell (1915) "On the Experience of Time," *Monist* 24: 212–33. J. M. E. McTaggart (1968) *The Nature of Existence*, ed. C. D. Broad, Grosse Pointe, MI: Scholarly Press, vol. II (1908) "The Unreality of Time," *Mind* 18: 457–74. A. N. Prior (1967) "Changes in Events and Changes in Things," in *Time and Tense*, Oxford: Oxford University Press, 1–14, reprinted in Le Poidevin and MacBeath Murray (eds) (1993) Oxford: Oxford University Press, pp. 35–46.

4 C. D. Broad (1921) "Time," in J. Hastings (ed.) *Encyclopedia of Religion and Ethics*, New York: Scriber, vol. 12, pp. 334–45; the quoted passage occurs on p. 334. Bertrand Russell (1915).

GLOSSARY OF TERMS

A-series The series of positions which runs from the far past through the near past to the present, and then from the present through the near future to the far future.

B-series The series of positions which runs from earlier to later.

Date-analysis On this analysis, "I am happy now," uttered on December 1, 1993, is analyzed as "I am (tenselessly) happy on December 1, 1993."

Dialectical method This method appeals to basic principles, reasoning and argument to determine the logical implications of a theory and subsequently its truth.

Method of translatability Determining the reality of temporal properties or temporal relations by considering whether or not tensed discourse can be translated without loss of meaning into tenseless discourse or vice versa.

Phenomenological method Phenomenology concerns how the world is given to us. The phenomenological method argues for one view of reality as opposed to another by appealing to which view more closely fits the world as we experience it.

Proposition The meaning or informational content of a sentence. For example, "It is raining" and "Il pleut" both have the same meaning, although the sentences used to express that meaning (or proposition) differ.

Psychological analysis On this analysis, "The sun is shining now," is analyzed as "The sun's shining is perceived or simultaneous with an object of perception."

Static aspect of time Time viewed solely in terms of the temporal relations of simultaneity, earlier and later. When viewed in this way time is called "static" because events standing in these relations never change their position relative to one another.

Temporal becoming A phrase used to refer to the passage of time, or the passing of events through time, from the far future to the near future to the present, and from the present to the more and more distant past.

Token-reflexive analysis Attempts to analyze a tensed sentence such as "It is now 1994" in terms of a tenseless sentence which expresses a temporal relation between a sentence token and a certain fact. Thus, "It is now 1994" is analyzed as "'It is now 1994' is simultaneous with (some temporal part of) 1994."

Transitory aspect of time *See* temporal becoming.

Transitory temporal properties The non-relational properties of pastness, presentness and futurity.

Truth value The truth value of a proposition refers to its being true or its being false. In some logics, a proposition has an indeterminate truth value, i.e., it is neither true nor false.

STUDY QUESTIONS

1. Explain the differences between the tenseless and the tensed theories of time.
2. Why are the various attempts to translate tensed language by means of tenseless language inadequate, according to the tenser? How might a detenser reply?
3. What dialectical difficulty does Ivan raise against the tensed theory of time? How does Sophia defend the tensed theory against Ivan's criticism?
4. What is Ivan's account of change?
5. What phenomenological difficulty does Sophia raise against the tenseless theory of time? How does Ivan respond to Sophia's objection?
6. Which theory of time do you think is preferable? Why?

FURTHER READING

Broad, C. D. (1921) "Time," in J. Hastings (ed.) *Encyclopedia of Religion and Ethics*, New York, Scribner.
A good presentation of the tenseless theory of time by a philosopher who later came to adopt a tensed theory.
Kiernan–Lewis, J. D. (1994) "The Rediscovery of Tense – A Reply to Oaklander," *Philosophy* 69, 265: 231–3.
Argues that our experience of ceasing to exist establishes that reality contains irreducibly tensed features.
Mellor, D. H. (1981) *Real Time*, Cambridge: Cambridge University Press.
Amongst the first to champion the view that the untranslatability of tensed sentences into tenseless ones is not an argument against the tenseless theory.
*Oaklander, L. Nathan (1984) *Temporal Relations and Temporal Becoming: A Defense of a Russellian Theory of Time*, Lanham, MD: University Press of America.
A systematic defense of the tenseless theory and critique of the tensed theory.
*— and Smith, Quentin (eds) (1994) *The New Theory of Time*, New Haven, CT: Yale University Press.
Contains selections written in the 1980s and 1990s pertaining to the tenser versus detenser debate.
Prior, A. N. (1959) "Thank Goodness It's Over," *Philosophy* 34: 11–17.
In this classic article Prior argues that tenseless theory cannot adequately explain the relief we experience when an unpleasant event is over.
*— (1968) *Time and Tense*, New York: Oxford University Press.
Schlesinger, George (1995) *Timely Topics*, Hampshire, UK: Macmillian Press Ltd.
Contains a novel account of the moving "now" and the passage of time.

Seddon, Keith (1987) *Time: A Philosophical Treatment*, New York: Croom Helm.
A clear introduction to the debate between tensers and detensers from a defender of tenseless time.
*Smith, Quentin (1993) *Language and Time*, New York: Oxford University Press.
The most sophisticated and elaborate defense of the tensed theory to date.
*Swinburne, Richard (1990) "Tensed Facts," *American Philosophical Quarterly* 27, 2: 117–30.
Contains a critique of Mellor's tenseless theory and an explication and defense of the tensed theory.

Dialogue 7

Personal identity

The cafeteria in the student union.

IVAN: Until now we have been mainly concerned with the role of time in questions concerning the nature and reality of change. Although there is certainly more that could be said on that issue I suggest we turn to another equally pressing issue, namely, the topic of *identity*. After all, the original problem that motivated our discussion today was change, and change involves identity.

SOPHIA: As I recall, questions about identity began to surface when Ivan claimed that a thing's changing involves no more and no less than its successive stages or phases exhibiting qualitative variation. I then raised the question of what constitutes the successive phases of *one* thing.

PHIL: Let me see if I understand the issue. A thing changes by having a property at one time and then losing it at another. Take the rose on that table over there. Common sense says it is one and the same rose that was once in bloom and fragrant, but is now dry and odorless. Indeed, how could it be *the rose* that changes unless it is the same rose? Now, given that change involves identity, our present task involves giving a coherent account of what constitutes the identity of something, say, a rose, undergoing change. Is that right?

SOPHIA: Yes it is.

ALICE: To put the issue in terms of roses is pleasant, but it lacks significance for me. The issue becomes more real, and more important to me, if I think about myself and my own identity. I am not what I used to be. Over the past ten years I have matured physically and emotionally. Over the next ten years probably all the matter of which my body is presently composed will be replaced, and my mental life will exhibit great changes. For all I know there will be changes in my personality too, for example, my approach to life, my beliefs, goals, aspirations and the like, may to some degree change. And yet, there is one life, my life, that endures throughout all this change. But what constitutes one life, my life? In what sense are the various and different stages of my life unified so as to constitute the stages of one and the same individual person, me?

SOPHIA: Those are good questions and it's a good idea to focus on personal identity for that topic is not a purely abstract issue of concern only to professional philosophers, but connects with important issues that matter to all of us. For example, most of us have thought about death and the question of what, if anything, is in store for us after death. However, to ask the question "Is survival of bodily death possible?" is virtually equivalent to asking the question "Is it possible for *me* to exist after my body ceases to exist?" But what am I, and in what does my identity through time consist? These are some of the philosophical questions of personal identity.

PHIL: I have always been attracted to Plato's answer to those questions. He thought that a person was composed of two parts: a physical or bodily part and a more fundamental immaterial or non-bodily part which he called the soul. He also maintained that the soul was the seat or subject of consciousness and the bearer of personal identity. Plato reasoned that my thoughts can change, my body can change and indeed, everything about me can change, but what is *invariable* is the inner me, what I am essentially, my soul. Thus, for Plato, our physical body has very little to do with our identity. Rather, on his view of personal identity, the unity or sameness of a life is founded upon the unity or sameness of the immaterial soul.

SOPHIA: Although I am inclined to be more corporeal and less spiritual when it comes to personal identity, I can think of two reasons why someone (though not I) might adopt the Platonic view. For one, our language suggests that there is one and the same subject of various mental states. Thus, for example, we say such things as "*I* am happy," "*I* am sad," *I* am thinking about personal identity" and "*I* am experiencing great joy at the success of my children." Certainly, so the argument goes, in each of these cases there is a single referent for the personal pronoun "I" and the soul serves the function of being that referent.

Consider also that experiences are never just floating around: if there is the feeling of pleasure, or the seeing of a movie, the playing of a game of cards or the experiencing of joy, then there is always a subject or owner of those states, that is, a person feeling the pleasure, seeing the movie, playing with the cards and experiencing the joy. Since whenever there is an experience there is a subject who has that experience, we can reasonably conclude that the subject or owner is none other than an immaterial substance or soul.

IVAN: I agree our language suggests a single subject for each of the different states of consciousness we successively possess, and I could hardly deny that experiences must be owned. However, I do not think that either truth implies the existence of a *single substance* that remains the same throughout our life; still less does it imply the existence of an immaterial soul.

SOPHIA: Really, why not? What are your arguments?

IVAN: I will explain them in a minute, but first I need to make some distinctions. You know somebody once said "Philosophy is the art of making distinctions," and there is much truth in that saying. Anyway, there are two fundamentally different approaches to our identity through time: the *substance* and the *relational* approach. Both approaches can agree to the points Sophia made, but disagree about the proper *analysis* or metaphysical *interpretation* of them. The Platonic view is an example of the substance analysis. On this analysis, judgments of personal identity, for example, "The individual who is discussing personal identity this afternoon is the same person as the individual who was discussing time this morning" are judgments about a single continuing entity, me, that is wholly present this morning and this afternoon. On the substance view, I *have* experiences, but these experiences are not *parts* of me. Rather they are exemplified or inhere in the substance which I am.

A picture might help to see what is involved. Suppose I am feeling joyous at one time and feeling not so joyous (because, say, I am angry) at another time. The substance view could picture the truth that underlies those facts as follows:

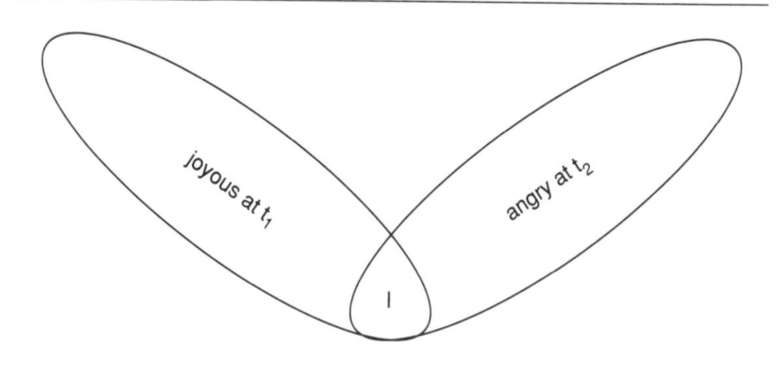

I = substance

On this view I persist as a unity throughout time, and my changing involves the inherence or exemplification by me of different properties at different times.

IVAN: On the relational view, on the other hand, the successive experiences that I have throughout my lifetime are not each related to an underlying substance or common unchanging constituent. Rather, when I judge that I am the same person today as I was yesterday, my judgment is true because the experience that I am having now and the experiences that I had yesterday fit together and are related in a way appropriate to constitute a single life.

ALICE: Do you mean that on the relational approach there is no intrinsically unchanging essence or core that I am?

IVAN: That's right. For the relationalist, instead of a substance having experiences, there are just the experiences themselves, and personal identity is nothing over and above the sum of the experiences *appropriately related*. Thus, my changing from being happy to being sad could be reflected in the following picture:

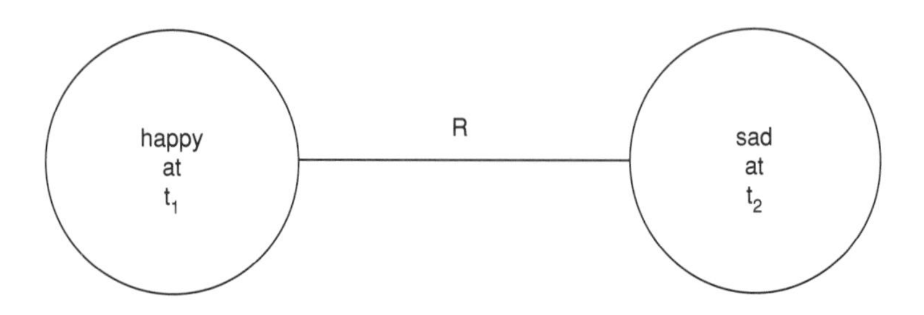

R = relation

ALICE: I see. On the substance theory, there is a substance to anchor or unite all thoughts, experiences and emotions which belong to one person. Whereas on the relational view there is some kind of string that ties or relates the different temporal segments into the phases of one life.

IVAN: Yes. On the relational view, "I" refers to the whole of which successive experiences are (temporal) parts.

PHIL: What you have not yet explained is how the facts Sophia appeals to, those about language and about the necessary ownership of experiences, can be accommodated on a view that rejects the existence of a substance.

IVAN: Well, on the relational view, to say that no experience is *unowned* is interpreted to mean that no experience exists in isolation from the whole of which it is a part. And to say that "*I* am happy at one time and *I* am sad at another" is analyzed in terms of two different experiences being *parts* of one and the same life. Thus, I don't see how the truisms Sophia appeals to can be used as reasons for believing in the existence of the soul, or any other kind of substance for that matter. To be fair, I should add that my argument doesn't prove the soul does not exist, although I must admit to having grave doubts about its existence.

ALICE: I suppose it is my religious upbringing, but I have always believed that I had a soul. I would search my soul to find answers to troubling questions, and I believed my most serious decisions arose from the soul, that is, from me. Why is it that *you* doubt the existence of the soul? And if the sameness of your soul does not constitute personal identity through time, then in what does personal identity consist?

SOPHIA: You would make an excellent philosopher Alice. You ask a lot of questions and never accept unargued-for assertions. That's a good trait, typical of a philosopher.

ALICE: Thank you. So tell me, Ivan, how would you respond to my questions?

IVAN: What I have to say on these issues is not original at all. David Hume denied the existence of the soul in the eighteenth century and his views have found sympathy with empirically minded philosophers ever since. I mean with philosophers who are guided by experiment and observation in determining what exists.

ALICE: What are his arguments?

IVAN: Hume argued that the limits of meaningful thought and language were delineated by experience. Thus, if something was neither experienced by sense perception or introspection, and if it could not be built up from what was experienced by the senses or introspection, then it did not exist and it was not meaningful to talk about it. He then claimed we do not have any experience of a soul or anything else that remains the same throughout our entire lifetime. Indeed, when we look into ourselves all we find are various emotions, feelings, thoughts, etc.; we do not find any simple substance that supports these "perceptions" (as Hume calls them) or in which they inhere. He concluded that since we do not experience the soul it is a mistake to suppose it exists or that we can intelligibly discourse about it.

ALICE: If Hume is right, then what is it that makes me me from one time to another? If there is no single entity or continuant that persists throughout the change of my perceptions; if there are just the changing perceptions or experiences themselves, then wherein lies my personal *identity*?

PHIL: Wasn't Hume the philosopher who said he could not come up with an answer to that question?

IVAN: Yes, as paradoxical as it might seem, Hume believed that there was no such thing as personal identity. In fact he didn't believe that there was non-personal or ordinary object identity either. He thought all identity was a philosophical fiction.

ALICE: Well, that's nice. Do you mean that *I* have not really been talking about personal identity for the past fifteen minutes?

IVAN: No, not really. In reality there have been many different Alices each with their own thoughts and emotions existing at successive minutes.

ALICE: Hume's theory seems implausible. Is that the only alternative to the Platonic theory of personal identity?

IVAN: No, the relational view may be able to account for your being you through time.

ALICE: What are some examples of the relational view?

IVAN: There is one that I have always liked, maybe because it enables me to make sense of life after death without committing me to the existence of a mysterious entity like the soul, and without committing me to reject what I strongly believe, namely, that after several more decades my body will cease to exist.

ALICE: Sounds too good to be true. I mean, given that my body will die, and I am not my soul, how is it possible for *me* to survive my bodily death? What is this new view of personal identity that you have in mind?

IVAN: It's a psychological approach that has a fancy name, non-branching psychological continuity! Quite a mouthful, but really very easy to understand until we get to the non-branching part. Let me develop it for you this way. Tell me, Alice, what are you doing now?

ALICE: What do you think I'm doing? I, or rather we, are talking and reflecting on the topic of personal identity.

IVAN: True enough. How do you know that is what we are doing?

ALICE: If you ask me to prove it to you, I won't because it's obvious. I am simply aware of what I am doing now, namely, talking to you.

IVAN: Fine. You are aware of our conversation as we are conversing. Philosophers sometimes refer to this kind of awareness as self-awareness or self-consciousness. This is the consciousness we have of ourselves when we are, say, perceiving, talking, thinking or whatever. I know what I am doing right now because I am conscious of it.

ALICE: So what does this have to do with personal identity?

IVAN: Well, the psychological approach emphasizes consciousness or, more accurately, self-consciousness, in its account of personal identity. On the psychological approach, what links a present stage of a person to a past stage of the same person is that the present stage contains a *memory* of the consciousness of the past stage. I can remember what you did yesterday in that I can remember having seen you do something in the past. What I cannot remember, however, is your experience or consciousness of having done that past action; only you can remember that. Only you can remember from the inside, from a first-person perspective, having done something in the past.

ALICE: Yes, and . . .

IVAN: It is just that psychological connection, memory, that unites the past you with the present you. In other words, if you can remember (from the inside) the consciousness of having acted in a certain way in the past, you are the person who in fact acted that way in the past. The existence or continued existence of the soul has nothing to do with it.

ALICE: But if my soul does not continue to exist, then how can I continue to exist? What am I if not my soul?

IVAN: You are a succession of experiences, including thoughts, feelings, emotions and actions, and you continue by having those experiences related in the right way. As far as the soul is concerned, John Locke, whose views on personal identity provide the starting point for psychological theories, thought that there was a soul, but wondered if perhaps he didn't have a different soul at every moment. He thought that even if he did change his soul he would still be who he is if he could remember who he was! Look at it this way. Suppose that I was on my deathbed and I died. Suppose that when I died my soul left and went to another place. Suppose further that a few minutes later a different soul (not my old soul) entered my body and I came alive again with the same body, same memories and same character as that which I had before I "died." I would say that I came back from the dead, or that the person I was before my old soul left is the same person as I now am with this new soul. But then, the soul I have does not determine my identity.

PHIL: Look at it from another direction. If my old soul continues to exist, without either my body or memories, then I see no reason to think that I continue to exist. So, even if my soul exists, and even if it continues to exist after bodily death, it does not follow that I survive bodily death, since I am not my soul. My body or my memories, and possibly both, are essential to whom I am.

SOPHIA: I have never been attracted to the Platonic view, but I don't see how Ivan's simple version of the memory theory can be the complete explanation of personal identity either.

ALICE: What are some of the problems with it?

SOPHIA: First of all, if personal identity is defined in terms of memory, then how can you explain the fact that we sometimes do what we cannot remember doing? Certainly, the fact that we forget things, even beyond the possibility of recall, does not mean that we did not do them.

IVAN: True enough, but Locke's view can be modified to accommodate that difficulty so long as we distinguish between memory connectedness and memory continuity. Connectedness involves a direct relation between the memory and the action or thought remembered. For example, I am memory connected to our conversation on time this morning, and this morning I was memory connected to my having gone to bed the night before. Now, however, I am not memory connected to my public school graduation. Since it happened so long ago I simply cannot remember it. Nevertheless, there is memory continuity between me now and me then; today I can remember what I did yesterday, and yesterday I could remember what I did the day before and so on back to the day I graduated from public school. Thus, we should define the view in terms of memory *continuity*.

SOPHIA: That still does not deal with a second problem with the memory theory, namely, that memory is too restrictive to define personal identity; other aspects of a person's psychology need to be considered too. Thus, a person's personality, that is, his or her tastes, interests, beliefs, desires, intentions and values are also important and need to be included in any adequate and complete account of personal identity.

ALICE: All this is going very fast. Please, will someone put it all together for me? I change and yet I am one persisting person who undergoes the change. So what is the me that remains the same?

IVAN: Well, the psychological approach emphasizes the fact of change, and downplays the fact of identity. That is, what makes you you from one moment to the next is the continuity of your memory and your character from one moment to the next. By "continuity" I mean the slow and gradual change of your memories, your beliefs, your attitudes and more generally, your psychology.

SOPHIA: Admittedly, there is a great deal of plausibility to this developed view, but a problem that still bothers me about the psychological approach has to do with duplication or branching. I cannot be identical to two people and two people cannot be identical to me. But if the psychological approach is correct then either of these scenarios is possible.

ALICE: How so?

SOPHIA: Remember a few years ago there was the Jarvic 7, an artificial heart? Even today, people are on dialysis machines which are, in effect, artificial kidneys. Well, suppose that sometime we become proficient in producing artificial brains; machine-like devices that are able to function as the human brain functions, preserving all the memories and personal characteristics associated with the original person's brain.

Then we could survive our bodily death by getting a new brain since that would preserve our psychological continuity, and therefore, it would preserve our personal identity too.

PHIL: A similar case would occur if sometime in the future there exists a Star Trek form of space travel. One philosopher I've been reading lately, Derek Parfit, describes how such exotic space travel (which he calls "tele-transportation") might take place. He says that

> a scanner here on Earth will destroy my brain and body, while recording the exact states of all my cells. It will then transmit this information by radio. Traveling at the speed of light, the message will take three minutes to reach the Replicator on Mars. This will create, out of new matter, a brain and body exactly like mine. It will be in this body that I shall wake up.[1]

On Parfit's view, the individual who entered the cubicle and the individual who awoke on Mars are one and the same person. The replica on Mars is psychologically continuous with me and so, given that personal identity consists in psychological continuity, is me.

SOPHIA: All well and good, but what then would happen if the machine operator forgot that he transported me to Mars and transported me a second time so that there are *two* future persons who are psychologically continuous with me? Or suppose, to return to my thought-experiment, a second artificial brain, just like mine, was constructed and put in a different body before I was dead. Then, given the psychological approach, there would be two different mes and that is absurd.

ALICE: Actually, that doesn't sound all that bad. For one thing, if I become two people, then I (or at least one of me) will be able to catch up on some reading, and I (or at least one of me) can attend to other projects for which I haven't been able to find the time.

IVAN: Yeah, sure. And suppose the "two" of you become separated, and at some much later time met for a game of tennis. Surely, if they are both you, then they are the same person, for you are one person and not two. Thus, if I were watching the match I would have to say that in the case described, there is one person playing tennis with herself! Rather silly, don't you think?

ALICE: I suppose so.

IVAN: At this stage of the argument the non-branching clause is introduced to prevent absurdity. The idea is that if there are two people psychologically continuous with you then your identity is not preserved, but if there is only one such person then you retain your identity. In other words, one way of avoiding the duplication or branching paradox is to claim that psychological continuity doesn't preserve identity when there are two people involved. Hence the analysis of personal identity in terms of *non-branching* psychological continuity.

SOPHIA: That strikes me as *ad hoc*; a view conjured up just to solve a problem. Moreover, it makes the question of your identity dependent on matters that are extrinsic or external to you. Dependent, for example, on whether or not a duplicate is made of you, and that is rather counter-intuitive. What seems more likely is that if psychological continuity doesn't give us identity when there are two people involved then it doesn't work when there is only one.

ALICE: Great, I am not my mind, understood as the unchanging me that persists through time, and I am not my mind, understood as my changing psychological characteristics appropriately related. Well, then, what am I?

SOPHIA: I think you two are playing around with two different questions about personal identity. The first is: what is our concept of ourselves? This is not so much the question of "*What* am I?" but more like the question "*Who* am I? asked by an adolescent going through an identity crisis. The second is: what is the nature of personal identity through time? It may turn out that these two questions – one psychological, the other philosophical – are more closely connected than most philosophers have previously thought. I am not sure. As far as my own views are concerned, neither the Platonic view nor the psychological approach are convincing since I have always been partial to a physically based approach.

ALICE: Could you explain that to me?

SOPHIA: In its crude and unsophisticated form, the physical approach maintains that I am my (live) body, and that the continued existence of my (live) body constitutes the continued existence of me.

IVAN: Before I raise some questions about your account, I want to clarify what I have been saying. The psychological approach does not dispense with the body entirely. A person's character could hardly be manifested if he or she didn't have a body, and it is difficult to comprehend how a person could have memories without having a brain. However, while having *a* body or *some* body is, on any plausible version of the psychological view, necessary for personal identity, what is not necessary is that I have *this particular body*. Any body structurally similar would do. Thus, the psychological approach downplays the importance not of the body *per se*, but of bodily identity in matters of personal identity.

SOPHIA: Excuse me Ivan, but before you question what I have to say let me offer a further objection against your view since it will help to clarify my own view. It seems to me that your view really does need to incorporate the notion of bodily identity if it is to be useful.

IVAN: What do you mean?

SOPHIA: Take the concept of memory, which is included in your account of personal identity. Suppose that two people (who are room-mates) seem to remember from the inside turning off the stove before leaving the house. Of course, only one of the room-mates actually turned off the stove and so only one of them really remembers doing so. Who is it? Alternatively, what is the difference between the person who really remembers and the person who only seems to remember?

ALICE: The person who really remembers is identical with the person who turned off the stove.

SOPHIA: Exactly. Memory, which on the psychological view, helps to define personal identity, incorporates the concept of identity. More specifically, the person who really remembers is the person whose body is identical with the person who turned off the stove. Similarly, personality characteristics, for example, attitudes, beliefs, biases, hopes, intentions, etc., are characteristics that more than one person can share. After all, we do talk of personality types. Thus, as in the case of memory, if you and I have exactly similar continuous beliefs, in order to distinguish my continuous belief from yours, a body must be introduced and not just any body, but my particular body.

IVAN: I see what you are getting at, and I realize that more has to be done to avoid appealing to an unanalyzed sense of identity in my psychological account of personal identity. Still I am not comfortable with the bodily view.

SOPHIA: What do you find objectionable?

IVAN: You say that personal identity requires bodily identity, and that the continued existence of the person involves the continued existence of the person's body. But my whole body need not continue to exist in order for me to exist, since I can be in a car accident and lose part of my body, but still retain my personal identity. Or, I can have a heart transplant, in which case my whole body does not continue to exist, although I do. At most it seems that on the physical approach enough of my brain must continue to exist in order for me to continue to exist, but I have problems with that too. The importance of *my* brain for my personal identity seems to be nil. I have never seen my brain, and don't care to. I am attached to my brain, but only because it somehow is the house, or storage place, and causal basis for my psychology. If I was able to get another brain that could sustain my memories and character (and perform the brain's other functions) as well as this particular hunk of matter, then it would not at all matter to me that I lost my brain through surgery. So my point is that neither this particular body nor this particular brain is necessary for my personal identity. In short, while having *a* body is essential to my personal identity, having *this particular body*, in part or whole, is not.

SOPHIA: I am not sure that I agree with you on either point. You argue that personal identity cannot require bodily identity because our bodies can lose their identities, that is, lose a part or exchange one part for another, while our personal identity can remain intact.

IVAN: Yes, that's one of my arguments.

SOPHIA: I thought so, but I don't think it is a very good one. Admittedly, in the case of a heart or liver transplant, all the parts of my original body do not exist, but my body can change its parts and still be the same body. And even if we modify the physical approach so that only the brain, or some part of it, is required for personal identity, the importance of its continued existence should not be underestimated.

PHIL: Excuse me, professors. The discussion is fascinating, but I think it is getting off the track. We are concerned with identity, so the appeal to bodily identity or brain identity as the basis for solving the problem of personal identity leaves the basic issue that we have been addressing, "What constitutes identity through change?" untouched. Sophia's view leaves me wondering what constitutes bodily or brain identity.

ALICE: I agree with Phil. Important identity questions still remain on the physical approach. For example, Sophia says that the body can change its parts and still remain the same body. Well, if it has different parts then what makes it the same?

SOPHIA: There are many different answers to that question, but rather than try to defend one of them, I want only to review the two broad camps into which most of the answers fall. As in the case of personal identity, the responses to the question of bodily identity are either relational or substance views.

ALICE: Could you go over the distinction one more time?

SOPHIA: Certainly. Before we start doing any philosophy we can agree, commonsensically, that my body like yours is a continuant; it persists or exists at more than one (or through) time. The philosophical issue concerns the analysis of that truth. What is a continuant and how does a continuant persist through time?

PHIL: Yes, those are the questions, now what are the answers?

SOPHIA: The relational account, which incorporates the notion of temporal parts, is one answer that has found favor among many contemporary philosophers.

ALICE: What are temporal parts?

SOPHIA: It's a bit tricky, but talking about spatial parts can give you the idea, I hope. [*Picking up a napkin*] Suppose we call this napkin "A." Now I'll draw a line through the middle and color one part, call it "Al," blue and leave the other part, call it "A2," white. Then we can say that the napkin A which exists in part at Al is the same as the napkin A which exists in part at A2, that it is one and the same napkin that is blue and white even though Al is not the same as A2.

ALICE: I see that, but those are spatial parts, what do you mean by temporal parts?

SOPHIA: Temporal parts are analogous to spatial parts. At different times of a thing's existence it is composed of different temporal parts. In other words, individual things, like my body, are *extended* through time as they are extended through space: they persist by having different, and suitably related, temporal parts at different times.

IVAN: To say that one and the same body changes is to say that the entire extended whole has a temporal part with one property (say, being composed of certain physical parts) and that another temporal part has a different property (say, being composed of different physical parts) and those two temporal parts are so related as to constitute one and the same body. At this point the various versions of this approach are distinguished in terms of the "appropriate" relation or tie that "glues" the temporal parts together, with spatio-temporal continuity and causality being the leading competitors.

PHIL: Evidently, the relational view denies that there is any deep way in which a body at two different times is identical. Identity is not founded upon a common constituent of the successive stages or temporal parts of a persisting thing, but is built up, so to speak, from the relations between those parts or stages which are themselves in no way identical to each other.

SOPHIA: That's the idea. On the other hand, for the defender of substance there must be something that literally endures through time that is the ultimate ground and basis for the identity of the changing thing; a relation that connects different things is not enough. For the substantialist, the entire object, or at least some component parts of it, is wholly present at more than one time. Thus, when a body is changing its properties there must be at least some of the original matter or stuff of which it is composed, both before and after the change, in order for it to be the same changing body.

IVAN: The issue of the substance versus the relational analysis of identity through change is fundamental since it arises whether we are discussing the topic of personal or non-personal identity. In the light of that I have an idea. I have to get back to the office since I have a class in a few minutes, but after my class perhaps we can explore some of the connections, if any, between the substance and relational accounts of identity on the one hand, and the tensed and tenseless accounts of time on the other.

PHIL: I am liable to overdose on philosophy, but what the heck, let's go for it.

ALICE: Sounds good to me.

SOPHIA: We'll meet back at the Philosophy Department at 3:30.

NOTE

1 Derek Parfit (1986), *Reasons and Persons*, New York: Oxford University Press, p. 200.

GLOSSARY OF TERMS

Continuant An entity that remains the same through time and change.

First-person memory The memory of the consciousness of having done something as opposed to the memory of something being done.

Non-branching psychological continuity Personal identity is defined in terms of psychological continuity in the 1:1 case. Thus, if two people, B and C, are psychologically continuous with person A, then there is branching and neither B nor C is identical with A.

Physical view of personal identity In its strongest form a person is simply a live human body and personal identity is defined in terms of bodily identity. A weak version of the physical approach would maintain that bodily identity is a necessary condition of personal identity.

Psychological connectedness Direct psychological relations, e.g., memory, intention, anticipation, between a psychological attitude and its object.

Psychological continuity A chain of direct psychological relations.

Relational view of personal identity Judgments of personal identity are based upon relations between the simultaneous and successive experiences that constitute the different stages of a person's life.

Self-awareness The consciousness we have of our experiences at the time we are having them.

Self-consciousness *See* self-awareness.

Soul The immaterial or non-bodily part of a person that is the subject of consciousness and the bearer of personal identity.

Substance view of personal identity Judgments of personal identity are based upon the existence of a single continuing entity that is wholly present at each moment of a person's existence.

Temporal parts Those who believe in temporal parts believe that anything that has a duration can be divided into successive and numerically different temporal stages.

STUDY QUESTIONS

1 What reasons are there for believing that I am a substance (or soul)? Are they good reasons?
2 How does Hume argue against the substance view of personal identity? What questions arise on Hume's own view?
3 What is the memory theory of personal identity? What are some of the difficulties with it? Can any of those difficulties be resolved? How?
4 What is the duplication problem? Do you think that Parfit's response to it is acceptable?
5 If your brain was transferred to somebody else's body and their brain was transferred to yours who and where would you be?

FURTHER READING

*Brennen, Andrew (1988) *Conditions of Identity*, New York: Oxford University Press.

A clear and far-ranging treatment of the concepts of identity and survival.

Kolak, Daniel and Martin, Raymond (eds) (1991) *Self and Identity*, New York: Macmillan Publishing Co.

A useful anthology containing articles devoted, in part, to the topics discussed in this dialogue.

Noonan, Harold (1989) *Personal Identity*, New York: Routledge and Kegan Paul.

An excellent summary and critique of various classic and contemporary theories of personal identity, as well as a statement of Noonan's own theory.

*Oaklander, L. Nathan (1987) "Parfit, Circularity and the Unity of Consciousness," *Mind* 96: 525–9.

Oaklander argues that Parfit's account of the unity of consciousness at a moment is open to the charge of circularity.

Parfit, Derek (1986) *Reasons and Persons*, New York: Oxford University Press.

A tremendously influential work defending the non-branching psychological continuity view of personal identity, and de-emphasizing the importance of identity in questions of survival and responsibility.

*Rovane, Carol (1990) "Branching Self-Consciousness," *The Philosophical Review* 99: 355–95.

This article is concerned with the question: can we preserve the first-person character not just of memory, but of forward-looking psychological attitudes, when we try to image the inner lives of branching persons?

Schechtman, Marya (1990) "Personhood and Personal Identity," *Journal of Philosophy* 87: 71–92.

Schechtman argues that psychological continuity theories are inherently circular and that discussions of personal identity should focus more on self-knowledge.

Shoemaker, Sydney and Swinburne, Richard (1984) *Personal Identity*, Oxford: Blackwell.

In this debate Swinburne adopts a dualist approach and bases personal identity on the sameness of an immaterial substance, whereas Shoemaker adopts a version of the psychological continuity view.

*Unger, Peter (1991) *Identity, Consciousness and Value*, New York: Oxford University Press.

Unger argues that a physically continuous realizer of our core psychology is necessary for personal identity.

*van Inwagen, Peter (1990) *Material Beings*, Ithaca, NY: Cornell University Press.

In this fascinating and provocative book, van Inwagen argues that there are no tables or chairs or any other visible objects except living organisms, and that the identity of a person through change depends on, and only on, the brain.

Dialogue 8

Personal identity and time

IVAN: We ended lunch by raising an issue that has not received much attention in philosophical circles, but which is an important issue that may be relevant to our discussion of time, change and reality.

ALICE: Could you please refresh my memory. I am still bemused by our discussion of the soul, the memory theory and personal identity.

IVAN: Sure. Over the course of the day we have come to see that questions of *change* are intimately and inextricably associated with questions concerning *time* and *identity*. Our discussion of time revealed that one fundamental issue concerns the correct description of time's passing. Is temporal passage metaphysically real, founded upon the changing or moving present, or is it really a myth whose metaphysical basis is our changing psychological attitudes toward a tenselessly existing string of events?

SOPHIA: And at lunch we distinguished two fundamentally different conceptions of identity through time. On the first, which we called the substance view, what grounds the persistence of a thing, such as person, is a substance or enduring particular without temporal parts. On this view, change consists in one and the same substance having or exemplifying different properties at different times. On the relational view a temporally whole object (from birth to death) is a succession of temporal parts, and persistence through time is based upon some relation or relations which unite the different temporal parts into a single temporal whole. Change is understood in terms of different temporal parts exemplifying different properties.

ALICE: Yes, I remember those pairs of theories. Now, what exactly is the issue you and Sophia wish to discuss today?

IVAN: Today we would like to explore what connections there are, if any, between the tensed and tenseless theories of time, on the one hand, and the substance and relational views of identity on the other. Is the tensed view (the view that only the present exists) compatible with the doctrine of temporal parts? Is the tenseless view compatible with the substance view? These questions strike me as interesting on their own account, but they might also be useful in our concerns. For if, as Alice and Sophia seem to think, the tensed theory of time is the only way to go, then any incompatibility between that view and the temporal parts approach will be relevant to their account of identity through change in general and personal identity in particular.

ALICE: I see what you mean, but how do you propose that we proceed?

SOPHIA: One way to explore the issue is to consider why someone might think that the doctrine of temporal parts is incompatible with the tensed view of time.

IVAN: Actually, I think that the two views are incompatible.

SOPHIA: Why is that?

IVAN: On the tensed theory of time, the locution "X moves through time" implies that X is present at different times. So, as the present moves from one time to the next (or as different events and moments come into existence), X allegedly moves along with it.

SOPHIA: Yes, that's right.

IVAN: The problem I have is this. Since X is a whole of temporal parts, *it* does not strictly speaking exist at different times – only its temporal parts do (different temporal parts at different times). If, however, X does not exist at different times, then X cannot be present at different times, and therefore X cannot move from one time to another. Thus, the temporal parts view of personal identity is incompatible with the notion that persons move through time.

ALICE: Could you go through that one more time?

IVAN: Look at it this way. Though different temporal parts of X can become present at different times, X cannot move from one time to another since X (as a temporal whole) does not literally exist at any time at which its temporal parts do. At best, each temporal part of X is successively becoming present, but it is difficult to understand how that would constitute *X* moving through time.

PHIL: I think I get it. On the tensed theory only the present exists. Thus, there is no sense in which the past and future parts of a whole exist now. But if only one part of a whole person exists at any one time then it is impossible to say that the whole person exists at any one time. Consequently, the person cannot move from one time to another.

SOPHIA: That is a reasonable argument against the compatibility between the tensed theory of time and the doctrine of temporal parts, but it is not conclusive. It may be objected that a whole can exist even if only one part is present, for it is possible that a whole has only one part. Thus, we could say that when time t_1 is present, the whole person X is composed only of the part P_1 that is then present, and when the later time t_2 is present, X is composed of the part P_2 that is then present. The same whole is composed of different parts at different times, and as the successive parts become present, come into existence, the whole moves from one moment to another.

IVAN: I agree that if a whole had only one part then it would be present if the part was present, and so it could move with the passage of time. However, to suppose that a whole has only one part creates a dilemma.

ALICE: Could you explain that to me?

IVAN: The whole is either identical with that part or it is not. If the relation between a part and the whole of which it is a part is identity, then a change in a part would imply a change in the whole. In that case, the coming into existence of a new and different temporal part would not help explain the movement of *one and the same person* from one time to another. If, on the other hand, a whole with only one part is not identical with that part, then what is it identical with?

PHIL: Maybe the whole is a substance.

ALICE: That won't do because a substance does not have temporal parts and so its existence is irrelevant to the question of whether or not the doctrine of temporal parts is compatible with the passage of time.

IVAN: Good point, Alice.

PHIL: Perhaps we can think of the whole person as an aggregate of different temporal parts that successively come into existence.

IVAN: Unfortunately, that alternative is equally unattractive. For if a whole is something distinct from its presently existing part and is the aggregate of all its parts, then it does not exist at any one time and so cannot, in any literal sense, move from one moment to the next.

SOPHIA: You mention a possibility that intrigues me. Instead of thinking of the tensed theory as maintaining that only the present exists, perhaps we ought to give reality to the past, present and future. We can think of time as being spread out like a string of beads, with all events existing, but add to the string the temporal distinctions of past, present and future. Then we could think of things moving through time in this way: "I" in the sense of a whole of temporal parts, am moving through time as successive temporal parts of me become present. First I am at $time_1$ when a temporal part of me, P_1, and t_1 are present, then "I" move on to $time_2$ when P_2 and t_2 become present, and this movement continues until I finally meet some future event E when P_3 and t_3 become present.

IVAN: I don't really see how this helps. Of course at any one time all the parts of X have some transitory temporal property; each one is either past, present or future. But then the most we can say of X is that it is *partly past, partly present* and *partly future*. We cannot say of X that it is present *simpliciter*, but that is what we must be able to say for it to be true that "X moves in time" as presentness moves from one time to another.

PHIL: I am not following you. Could you give me another example?

IVAN: My point can be made most clearly by comparing the temporal case with the spatial case. A spatial whole which is part green and part red is, strictly speaking, neither green nor red, only partly each. Analogously, a temporal whole that is part past, part present and part future is, strictly speaking neither past, present nor future, but only partly each. The problem, then, is that the entire object cannot be said to move through time as presentness moves from one time to another since it does not exist at any one time.

SOPHIA: But I do not accept your analogy between space and time. Unlike the spatial case, in the temporal case a whole may have the same temporal property that a part has even if other parts of that whole exemplify different temporal properties. In other words, I believe that *a temporal whole W is present if one of*

W's temporal parts is present on the grounds that its denial entails a contradiction. Thus, if W is past (or future) when one of its temporal parts is present, then W is past (or future) when its earliest (or latest) temporal part is present, which is absurd.

IVAN: The problem with your argument is that it proves too much, since analogous reasoning can be used to support the thesis that W is past (or future) if one of its temporal parts is past (or future). For example, to suppose that W is present (or future) if one of its temporal parts is past, entails the absurdity that W is present (or future) when its latest temporal part is past. So, I don't see how we can infer that a temporal whole is present if one of its parts is present.

SOPHIA: I follow your argument, but I find the conclusion counterintuitive. Intuitively, we all agree that John is present if one of his temporal parts is present, that the earth is present if one of its temporal parts, say the 1988 part, is present and that World War II is present if one of its temporal parts, say the Battle of the Bulge, is present. Thus, I still suggest that as presentness moves from one temporal part of me to another, I move along with it.

IVAN: I don't think that our intuitions are as sophisticated as you take them to be. After all, do any non-philosophers really think that we have temporal parts? I don't think so. Perhaps we think that if a stage in a person's history is present, then the person is present. If, for example, I am mowing the lawn at one point in my life's history, then I am present or exist at that point in time. Agreed. But to talk about the history of a person is to assume that I am an enduring substance or continuant and not a perduring whole of temporal parts. An *enduring* substance is wholly present at each moment of its existence, and cannot be divided into temporal parts, although its history can. A *perduring* whole on the other hand, is not wholly present at each moment of its existence since it is composed of temporal parts which each exist at a different moment of time. The identity of a perduring whole is founded upon some relation that its temporal parts have to each other.

SOPHIA: You still haven't convinced me that the tensed theory of time is incompatible with the temporal parts analysis of personal identity. But even if you are right, your reasoning does nothing to undermine the tensed theory itself. For the tensed theory of time is compatible with the substance view of personal identity.

IVAN: What is your argument? How can the substance view explain the sense in which I am moving through time, given the tensed theory?

SOPHIA: Suppose an event E is three minutes in the future and a substance S is present. Then E comes to meet S in the sense that S remains present while E goes from being three minutes in the future to two minutes, one minute and finally to being present, at which time it meets S.

PHIL: How exactly does this analysis satisfy the definition of moving temporally or spatially closer to something?

SOPHIA: Well, something X moves closer to something Y if and only if the units of temporal or spatial distance separating X and Y decrease with the lapse of time. Suppose E is located at t_3 and S at t_0, which is present. If S persists from one present moment to the next, then as t_1 becomes present S passes from t_0 to t_1, and as t_2 becomes present S passes from t_1 to t_2. The number of temporal units' separation E from S decreases as time lapses, from three units (when t_0 is present) to two units (when t_1 is present) to one unit (when t_2 is present) and finally to zero units (when t_3 is present).

IVAN: Perhaps, then, the most sensible thing to do, if you want to maintain the tensed theory of time, is to hold the substance view of personal identity.

SOPHIA: I am not convinced about that, but will let the point stand.

IVAN: Having explored one connection, or lack of such, between the tensed theory of time and the temporal parts analysis of personal identity, let's turn to another pair of views, namely, the substance view of personal identity and the tenseless view of time.

SOPHIA: I can think of one argument for showing the incompatibility of the two views in question. It goes like this. If a person is an enduring substance S that exists tenselessly through time, and it changes, say, from being sober to being drunk, then it is true that S is (tenselessly) drunk and is (tenselessly) sober. Since that is a contradiction, if one is to hold the tenseless view, then one must claim that a person is a whole of temporal parts, for then being sober would belong to one temporal part and being drunk to another and the contradiction would vanish.

IVAN: This is a rather feeble argument since what is missing from your account of change is time! The difficulty of a thing having incompatible properties is resolved by the tenseless theory by introducing substantival time in the form of moments. The individual that exists at t_1 is the same as the individual that exists at t_2 even though it has different properties, since S has the property of being (tenselessly) sober-at-t_1 and drunk-at-t_2, and those different properties are compatible.

SOPHIA: I suppose you are right, although my problem with your "way out" concerns whether or not such a response really yields change.

IVAN: That is the issue we discussed this morning. Are there any other difficulties you see in this account of change?

SOPHIA: In fact there is one. Your assumption of the substantival rather than the relational theory of time is not required by the tenseless theory to solve the incompatible properties problem.

IVAN: Why not?

SOPHIA: On the relational theory, times are not moments, but sets of events. Thus, we can form the collection of all events simultaneous with any particular event and identify a time with this collection. Given this relational theory, "At t_1" in the clause "At t_1 P is thinking of Plato and at t_2 P is not thinking of Plato (but thinking of Descartes)" is analyzed as "simultaneously with the events in the set S_2 of simultaneous events."

IVAN: I don't think we can adopt the relational view if individuals are continuants and time is tenseless. To see why consider a world in which a substance P has always been thinking of Plato and Descartes in succession. That is, assume that events of the type *P is thinking of Plato* and *P is thinking of Descartes* have been recurring forever. In such a world how could we individuate one occurrence of thinking of Plato from another? Assuming that time is relational and tenseless, neither their spatio-temporal relations nor their constituents would be able to distinguish the different occurrences of either type of event.

SOPHIA: What then is to be done?

IVAN: One possibility is to introduce the tensed properties of pastness, presentness and futurity and individuate the different occurrences of the same event-type by means of them.

SOPHIA: That way out does not work for two reasons. First, since each event has all three tensed properties (at different "times") it is far from clear that this move would individuate the events in question without presupposing substantival time.

IVAN: Yes, but a second possibility is to individuate qualitatively indistinguishable events by treating them as numerically different temporal parts of "continuants." The problem with this alternative, however, is that we have abandoned the view of the self as a whole without temporal parts. Thus, it appears that on the view that persons are enduring substances and that time is tenseless, the only way to individuate different occurrences of the same event-type is by an appeal to substantival time.

SOPHIA: To this it may be objected that the occurrences in question could be individuated by reference to other events, such as brain states, to which they are related.

IVAN: I don't think that helps. For suppose we assume the intelligibility of Nietzsche's doctrine of eternal recurrence, according to which each event in the history of the universe has occurred and will occur an infinite number of times. The individuation problem then consists in distinguishing one entire state of the universe from another occurrence of that same state.

SOPHIA: OK, but now I see what is wrong with your view. On the substantialist conception of personal identity and the relational analysis of time, the same event cannot recur so that in the situation you are envisioning there are only two events, *P's thinking of Plato* and *P's thinking of Descartes*. After all, given the necessary truth known as the identity of indiscernibles (which states that if two things have all their properties in common then there is really only one thing), there would be nothing to distinguish the allegedly different occurrences of P's thinking of Descartes and P's thinking of Plato, and so there would only be the two events. Consequently, since there are no different occurrences of the same event-type in the world you envision, there is no need to introduce substantival time to individuate them.

IVAN: I agree that recurrence is not possible if we accept the literal continuant view of personal identity, the relational theory of time, the tenseless theory of time and the identity of indiscernibles. As I see it, however, that constitutes an indirect proof of my claim that on the tenseless view, continuants presuppose substantival time. For although I do have some difficulty in conceiving of a world in which all that existed was a disembodied mind that successively thought of Plato and Descartes forever, I can conceive of a universe that recurs eternally. I do not believe in either of these suppositions, but I do believe they are possible. Therefore, without denying the principle of the identity of indiscernibles, literal continuants or tenseless time, we can solve the problem of individuation only by assuming the existence of substantival time. What individuates different occurrences of P's thinking of Plato is that each occurrence is related to a different moment of substantival time.

SOPHIA: If you are right, then you have found an interesting connection between personal identity and time: the conjunction of the substance view of identity and the tenseless theory of time imply that time is substantival.

GLOSSARY OF TERMS

Enduring An enduring entity does not have temporal parts, and is wholly present at each moment that it exists.

Identity of indiscernibles A principle which asserts that if two things have all their properties in common then they are identical.

Perduring An entity is perduring if it is a whole composed of different and suitably related parts, such that each of these parts exists at a different time and no one of them endures from one time to another.

STUDY QUESTIONS

1 How does Ivan argue that the doctrine of temporal parts and the relational view of personal identity are incompatible with the tensed view of time?
2 Is the tensed view compatible with the substance view of personal identity? Explain.
3 Is there a fallacy in Sophia's argument for the incompatibility of the substance view of personal identity and the tenseless theory of time?
4 Evaluate Ivan's argument for the conclusion that the substance view of identity and the tenseless view of time together imply that time is substantival.

5 Does the fact that an entity has a duration imply that it has temporal parts?

FURTHER READING

Hoy, Ronald C. (1978) "Becoming and Persons," *Philosophical Studies* 34: 269–80.
An important article in subsequent discussions of personal identity and time.
Lewis, Delmas (1986) "Persons, Morality, and Tenselessness," *Philosophy and Phenomenological Research* 47: 305–9.
Argues that the tenseless theory of time is mistaken because it implies a theory of personal identity according to which persons cannot be held responsible for their actions..
Oaklander, L. Nathan (1987–8) "Delmas Lewis on Persons and Responsibility: A Critique," *Philosophy Research Archives* 13: 181–7.
Contains a response to Delmas Lewis's article and a discussion of time and substance.
—— (1992) "Temporal Passage and Temporal Parts," *Noûs* 26: 79–84.
Considers various versions of the tensed theory and argues that they are each incompatible with the doctrine of temporal parts.
Schlesinger, George (1983) *Metaphysics*, Totowa, N J: Barnes and Noble.
A defense of the tensed theory. In this book Schlesinger discusses Smart's argument, in "Time and Becoming" (listed below), that the tensed theory cannot make sense of temporal passage if persons are wholes composed of temporal parts.
Smart, J. J. C. (1981) "Time and Becoming," in Peter van Inwagen (ed.) *Time and Cause*, Dordrecht: D. Reidel, 3–15.
Contains a critique of the tensed theory of time, and an argument for the incompatibility of tensed time and temporal parts.
Smith, Quentin (1993) "Personal Identity and Time," *Philosophia* 22, 1–2: 155–67.
Smith argues, contra Hoy, Oaklander and Smart, that the tensed theory is compatible with either the continuant or the temporal parts doctrine of personal identity.

Part III

The nature of freedom

L. Nathan Oaklander

Dialogue 9

Fatalism and tenseless time

STEVE [*a philosophy major*]: I have been listening to your conversation with great interest and I would like to raise an issue that I think tilts the balance of the debate in favor of some form of the tensed theory of time.

IVAN: What issue is that?

STEVE: I was reading a book on fatalism by Steven Cahn where he argues that the tenseless view of time is logically incompatible with free will.[1] He went on to make the rather startling claim that the detenser can avoid fatalism only if he or she abandons reason itself. Since I am unwilling either to give up reason or accept that our will is unfree, I must reject the tenseless view which leads to these unwanted consequences and adopt the tensed theory which I believe avoids them.

IVAN: As you know, I am sympathetic with the tenseless view, and I would like to try to defend it against the charge that it leads to fatalism. Perhaps a good way to proceed is for me to ask you a few questions.

STEVE: OK. I am prepared with some of the articles and books that I have been using in my research.

IVAN: First of all, what do you mean by "fatalism"?

STEVE: As I understand it, fatalism is the view that the laws of logic alone can prove that no person has free will. For the fatalist the future has the same necessity as the past. When we contemplate an event in the past, we may think of it with joy or remorse, but we can now do nothing to undo or prevent it. We have neither the power to prevent a past action from having happened, nor the power bring about a past action that did not happen. Similarly, if fatalism is true, then we can now do nothing to bring about an action that will not in fact occur, or prevent from happening an action that will occur. In short, there are no more alternatives for the future than there are for the past, since both the past and the future are fixed.

IVAN: So, according to fatalism we may seem to deliberate over two possible courses of action, but actually our deliberation is an illusion since there is only one course of action that could occur.

STEVE: Yes, that's what the fatalist maintains.

IVAN: Certainly, if the tenseless view is committed to fatalism that would be a serious objection, but before we consider the argument for that thesis, I have another question. You say that fatalism entails that no person has free will?

STEVE: Yes.

IVAN: What do you mean by "free will"?

STEVE: A person has free will if and only if he or she is free with regard to some actions, and a person is free with regard to some action A if and only if she has it "within her power" to perform A and she has it "within her power" to refrain from performing A.

IVAN: But when is an action "within one's power" to perform or not to perform?

STEVE: I can do no better than cite Richard Taylor's understanding when he says,

> to have an action within one's power entails at least that nothing causes one to perform the action and nothing causes one to refrain from the action. This is not to say, however, that the action is inexplicable. It is simply up to the person himself whether he performs it.[2]

His idea, with which I concur, seems to be that an action is free if and only if it is caused by the person or self; the substantial agent whose spontaneous act of creation lies outside the sphere of scientific predictability and external causality.

IVAN: So, fatalism is the view that logic implies that there do not now exist alternative possibilities for the future.

STEVE: Yes, and I should add that only if we reinterpret certain laws of logic can we prove that fatalism is false.

IVAN: Well, I am not concerned right now with disproving fatalism, but only with the question as to why you believe that the tenseless theory implies that free will does not exist.

STEVE: Before I explain the argument let me briefly describe the tenseless view as I understand it.

IVAN: Good idea.

STEVE: The essence of the tenseless theory is that time consists solely of the temporal relations of *succession* and *simultaneity* between events. In other words, for the detenser, there are no monadic properties of pastness, presentness and futurity that events don and doff with the passage of time. Linguistically, this view has been expressed by claiming that all temporal statements are B-statements (or "eternal sentences"), that is, statements whose truth or falsity is independent of the time of utterance or inscription.

In his much discussed article on fatalism, Donald C. Williams characterizes a view of the world that is frequently associated with the tenseless theory. He says,

> I do wish to defend the view of the world, or the manner of speaking about it, which treats the totality of being, of facts or of events, as spread out eternally in the dimension of time as well as the dimensions of space . . . there "exists" an eternal world total in which past and future events are as determinately located, characterized, and truly describable as are southern events and western events. This does not mean that "time is a dimension of space" . . . but it does mean that past, present, and future are ontologically on a level with one another and with west and south and are equally real.[3]

IVAN: I agree with this characterization, but I should add the notions of events or facts as "spread out eternally" and of the existence of an "eternal world total" are ambiguous and may give rise to trouble. But leaving that aside, what is the argument against Williams and other detensers?

STEVE: For the sake of clarity I shall state the overall structure of my argument against the detenser in the following four steps:

1 The tenseless theory of time implies the law of the excluded middle, according to which every proposition, including propositions about the future, are either true, or if not true are then false.
2 If the law of the excluded middle is true, then no person has free will.

3 Therefore, if the tenseless theory is true, then a logical law implies that no person has free will, that is, the tenseless view implies fatalism.
4 Since, however, fatalism is false, the tenseless theory of time must be false too.

IVAN: The argument is valid and in some sense the first premise is true. Thus, for example, according to the tenseless theory, it is either true or false that "We will continue this discussion for thirty minutes." So clearly, the argument stands or falls on whether or not the second premise is true. What then is the argument to establish that the law of the excluded middle entails that no person has free will?

STEVE: In order to answer that question I want to consider Aristotle's argument for fatalism. For Aristotle argues that if the law of the excluded middle is true, then all events are necessary, that is, no action is within one's power to prevent or bring about. Aristotle's argument may be stated in the following five steps:

1 Every proposition must be either true or false, e.g., either there will be a sea-fight or there will not be a sea-fight tomorrow.
2 Assume that one person affirms today that a sea-fight will occur tomorrow and another person denies it.
3 The statement of the one person corresponds with reality and that of the other does not.
4 But in that case it must *already* be true that a sea-fight will take place tomorrow, such that there is now no possibility that it might not, or else it must *already* be true that a sea-fight will not take place tomorrow, such that there is now no possibility that it might.
5 Thus, as Aristotle says, in either case nothing is or takes place fortuitously, either in the present or in the future, and there are no real alternatives; everything takes place of necessity and is fixed . . . for the meaning of the word "fortuitous" in regard to present or future event is that reality is so constituted that it may issue in either of two opposite direction.[4]

Since the logic of this argument is impeccable, it seems to me that the tenseless theory, which accepts the law of the excluded middle, is committed to fatalism.

IVAN: Formally, Aristotle's argument is valid, I agree, but as an argument against the detenser it is either unsound or question-begging. To see why this is so let us carefully examine it.

STEVE: Fine.

IVAN: Let us suppose, as the detenser must, that premise (1) is true; that today, at time t_1, one person affirms that a sea-fight will occur tomorrow (at t_2) and that another person denies this. Then, if truth is correspondence, then premise (3) follows: either "The sea-fight will occur tomorrow" is true or "It is not the case that the sea-fight will occur tomorrow" is true. Now it seems to me that two absolutely essential questions that are rarely raised in discussions concerning fatalism and time are, first, "What is the nature of the correspondence relation between propositions and that to which they are correlated?" and second, "What is the nature of the correlated facts or states of affairs to which true propositions correspond?" By considering these questions we can get clear on the relations between your reconstruction of Aristotle's argument for fatalism on the one hand, and the theories of time and truth that are assumed by it on the other. Let me give you my interpretation of Aristotle and you tell me if it fits your view.

STEVE: Go ahead.

IVAN: Suppose, then, that at t_1 it is true that a sea-fight will take place tomorrow. What, we must ask, is the state of affairs in virtue of which it is true? It seems clear that in the case of true propositions, correspondence is a relation between propositions and states of affairs. What is also clear, or should be if premise (4) is to make sense, is that if a proposition is *now true* (at t_1) then the state of affairs in virtue of which it is true must *now exist*. In other words, truth goes with existence and present truth with present existence. Therefore, if a proposition about the future is now true, then it corresponds to an existing state of affairs; and if a state of affairs exists then it is present. To put the point still differently, if a proposition about the future is now true, then there exists *at present*, an external correlate in virtue of which it is true.

I think I am being fair in treating this as an implicit assumption of the argument. Do you believe that the present truth of a future tense proposition involves the present existence of a future tensed state of affairs?

STEVE: Yes I do. If it is true that I will go to the movies, this implies that at present there is the state of affairs that *I will go to the movies*. For that reason, it is no longer within my power not to go to the movies. In other words, the truth of the proposition, "I will go to the movies" does not force me to go to the movies, but that proposition describes a state of affairs which does force me to go.

IVAN: Alright. Once you build into the fatalistic argument the claim that truth involves correspondence between propositions and states of affairs that exist at the (present) time when the proposition is expressed, then the argument becomes very forceful. For if it is now true that there will (or will not) be a sea-fight tomorrow, then there now *exists* the fact that *there will (or will not) be a sea-fight tomorrow*, but then I cannot prevent (or bring about) a sea-fight from occurring (or not occurring) tomorrow. More generally, I cannot prevent a state of affairs in the future because if it is already true today that I will bring about a certain state of affairs tomorrow, then the future tense state of affairs already exists today, and I cannot prevent tomorrow what already exists today.

ALICE: Could you go through that reasoning one more time?

IVAN: Sure. The core of the fatalist argument is to move from a future tense proposition's being already true, or true prior to the time at which the event allegedly takes place, to the conclusion that the conditions that guarantee the truth of the future tense proposition already exist; that they exist *before* the allegedly "future" event takes place. Thus, if a future tense proposition, for example, "A will occur," is already true, then *at present* there exists either (1) the fact *A will occur* or (2) a set of causally sufficient conditions which "force" it to come about that *A occurs*. In either case, I cannot prevent A from occurring and consequently my action is not free.

SOPHIA: As you know, I do not believe that the tenseless theory is true, but I do think that your interpretation of the fatalist argument against it is highly plausible. Moreover, explaining the fatalist argument in the way you have done makes clear that it is not simply the logical law of the excluded middle that implies that no person has free will, but rather it is that law coupled with a theory of temporal truth and a decidedly tensed theory of time. For the argument assumes that true propositions correspond to facts (or truths) that exist *in time*, and that what exists in time exists in the present. Given these assumptions the principle of the excluded middle does lead to the view that the future pre-exists in the present, and therefore that all our actions are fixed in advance. However, on the tenseless view, the two assumptions that give strength to the fatalist argument are false. Thus, for the detenser, Aristotle's argument is unsound, and if Steve simply insists on claiming that the key assumptions are true, he is begging the question.

STEVE: Ivan, could you explain these last points, and defend them?

IVAN: Consider the future tense sentence "There will be a sea-fight." According to the tenseless theory of time as I understand it, the fact in virtue of which that sentence is true will vary depending on the time at which it is uttered or inscribed. For example, if the sentence in question is asserted today, at t_1, then the state of affairs which exists if the sentence is true is that *a sea-fight occurs (tenselessly) later than* t_1. However, *that* state of affairs does not exist at the time at which the future tense sentence is uttered, and it does not exist at any later (or earlier) time. Of course something does exist today, at t_1, namely the inscription or utterance "There will be a sea-fight," and assuming the utterance is true, something does exist at a time later than t_1, namely, the sea-fight, but the whole state of affairs that *the sea-fight occurs later than* t_1, does not exist at t_1, does not exist at t_2, and does not exist at any earlier or later time. Indeed, it does not exist *in* time at all. For the detenser, states of affairs which contain temporal relations between events are *eternal* in the sense of existing *outside* the network of temporal relations, but not in the sense of existing (or persisting) throughout all of time. Consequently, the present truth of a future tense sentence does not imply that the state of affairs in virtue of which it is true "pre-exists" in the present, or to use Leibniz's image, that the present is big with the future. And furthermore, the present truth of sentences about the future does not imply either that events are "already laid up" or that we are just like puppets whose decisions and choices are mere illusions. In short, the law of the excluded middle does not imply that no person has free will.

PHIL: Let me see if I am following you. You are saying that when properly understood the tenseless theory denies that future tense sentences correspond to facts that exist *in time* and it denies that existence is a tensed notion, or synonymously, that to be is to be present. Since the fatalist argument depends on both these unargued assumptions, and since Steve employs that argument against the tenseless theory, it follows that he is begging the question at issue.

IVAN: That sums it up nicely.

PHIL: Two questions immediately come to mind. First, why do tensers assume a theory of temporal truth in an argument against detensers? Put differently, why are detensers so often accused of being fatalists? Second, could one legitimately argue that even if we do not assume a theory of temporal truth, the detenser's commitment to "eternal" states of affairs entails that no person has free will?

IVAN: Good questions. Before we sum up this interesting discussion let me try to briefly answer them.

Concerning the first question I must admit that detensers have made some ambiguous remarks that do suggest a theory of temporal truth that would lead to the consequences that they wish to avoid. Recall that Donald Williams speaks of the world as:

> the totality of being, of facts or of events . . . spread out *eternally* in the dimension of time as well as the dimension of space. Future events and past events are by no means present events, but in a clear and important sense they do exist, *now and forever*, as rounded and definite articles of the world's furniture.[5]

STEVE: To me, the claim that past and future events are "spread out eternally" has unavoidable spatial connotations, and the claim that past and future events exist "now and forever" suggests that such events are temporally permanent or persistent particulars that somehow co-exist simultaneously (or "all at once").

IVAN: I can see how Williams's infelicitous use of the ambiguous words "eternal" and "forever" could mislead you. For, according to some, to exist eternally$_1$ (or forever) means to exist from the beginning to the end of time, or if time does not have a beginning or an end, to exist eternally$_1$ means to exist throughout all of time. Now, if one understands the history of the world as being "eternal" in that sense then the totality of facts or events exists at *every moment*, and therefore, all truth has its basis in facts that exist *at present*, in which case the fatalist argument goes through.

STEVE: Are there other notions of eternal?

IVAN: Yes. There are two other conceptions of "eternal," one of which a consistent detenser should adopt. In a second sense, to be $eternal_2$ has meant to be "timeless" whereby a "timeless" entity is one that neither *contains* entities that are in time, nor *occupies* a temporal moment, nor *exemplifies* temporal relations or non-relational temporal properties. Thus, it makes no literal sense to say of an $eternal_2$ entity that it is past, present or future; that it is earlier or later than another entity or that it occupies or occurs at a moment of time. Thus, for example, Platonic universals, like Justice itself, or Beauty itself are $eternal_2$, and so are the relations that obtain between and among universals.

STEVE: What is the other notion of "eternal"?

IVAN: To be $eternal_3$ has also meant to exist apart from time, although not, as an $eternal_2$ entity is, entirely independent of it. Thus, an $eternal_3$ entity is one that neither occupies moments of time, nor exemplifies temporal relations, nor has monadic temporal properties inhering in it. Rather, an $eternal_3$ entity is related to time in the following way: *it is a whole which contains successive parts*. We could say that although an $eternal_3$ entity is not contained in time, time is contained in it. Thus, the fact that *World War II is later than World War I* is $eternal_3$ because although not itself in time (or a term of a temporal relation) it contains time (a temporal relation) as a constituent. This view of eternity gives some meaning to an aphorism I favor, that *time is timeless*, i.e., though time contains temporal relations, time does not exemplify them.

A good statement of this conception of "eternal" is stated by J. S. Mackenzie when he says:

> There is no time outside the process. Hence the process as a whole might be said to be eternal though every particular part in it has a place in time. The eternal thus conceived, would not be the timeless, but rather that which included the whole of time . . . The process as a whole, when we thus conceive it, is not in time, rather time is in the process. Time is simply the aspect of successiveness which the eternal process contains.[6]

Thus, when Williams and other detensers claim that "the totality of facts and events are spread out eternally" they should be interpreted as holding that the world total forms a whole which contains successive parts. Perhaps the frequent claim by tensers that the tenseless theory of time is fatalistic results from confusing the first and third meanings of "eternal."

STEVE: I still don't see how your view avoids denying that we have free will. Admittedly, if "X will occur" is true at time t, then it is an eternal fact that *X occurs later than time t*. Still, it appears to me that *that* fact necessitates that X occurs later than time t, or synonymously, that it is not within one's power to prevent X's occurrence at a time later than t. So, my question still remains: does the existence of $eternal_3$ facts or states of affairs imply that no person has free will?

IVAN: I don't think so. After all, an affirmative answer to that question does *not* gain support from the law of the excluded middle. The law necessitates that one of a pair of contradictory statements is true, but it does not imply that either "P" is necessarily true or "not-P" is necessarily true. If I go to the Philippines on December 15, 1993 then my present statement (on October 15, 1993) "I will go to the Philippines on December 15, 1993" is true, and if I do not go to the Philippines in December then my present statement "I will not go the Philippines on December 15, 1993" is true. Thus, which statement is true depends on whether or not the fact *My going to the Philippines occurs two months later than October 15, 1993* exists or does not exist, but whether or not that fact exists depends upon what I choose to do on December 15. In other words, it is my later decision that determines which of two contradictory statements about the future is true, since it is my later decision that determines what $eternal_3$ fact exists. Thus, the existence of $eternal_3$ facts is not incompatible with our having it "within our power" to bring about or prevent certain events. On the contrary, it is because we *do* have it "within our power" (because we do *choose* or do not *choose*) to bring about or prevent certain events that certain $eternal_3$ facts exist.

SOPHIA: I have just one question: if the fact *Your going to the Philippines on December 15 occurs later than October 15, 1993* exists eternally, how can it depend for its existence on something that occurs on December 15? How can an entity that does not exist in time depend on something that does exist in time?

IVAN: Good question. $Eternal_3$ facts are not timeless in the sense in which abstract universals, or relations between universals are timeless. True, eternal$_3$ facts do not exist in time (i.e., as terms of temporal relations), but they do contain temporal relations and their relata as constituents. Thus, the fact that I am going to the Philippines at a certain date *includes* my decision on December 15 to go to the Philippines. That decision is an event (or a temporal part of my life) that exists in time, and if it did not occur on December 15, then there would be no such fact as my going to the Philippines on December 15. For that reason, the truth of a future tense sentence does not entail the existence of a fact which in turn determines my choice. Rather, my choice in December determines what (tenseless) fact exists, and hence what future tense sentence is true.

PHIL: Ivan, let me see if I can summarize the discussion up to this point. There are two ontological interpretations of the law of the excluded middle that are consequences of how one conceives of time and truth. According to Aristotle, Steve and others, to affirm *at instant t* that either "P" is true or "not-P" is true then implies that either "P" or "not-P" has an external correlate that exists *in* time, that is, *at instant t*. According to detensers, on the other hand, to affirm, at instant t, that "P" is true or "not-P" is true, implies a certain fact exists (tenselessly) or does not exist (tenselessly) in virtue of which it is true or false, but it does not imply that the fact in question exists at the time of affirmation. Indeed, it does not imply that the fact in virtue of which a tensed sentence is true exists in time at all.

IVAN: Yes, that is right. Furthermore, if there exists (tenselessly) the fact *I am in the Philippines on December 17, 1993*, then at any time (before, after or during my trip) it is true to say "I am (tenselessly) in the Philippines on December 17, 1993." However, the permanent truth of that sentence does not imply "determinism" in either the causal or the block universe sense (i.e., in the sense that all events exist simultaneously). The principle of the excluded middle ensures that a sentence about the future is either true or false at the moment, but it does not entail that the causes of the future, or facts about the future, exist in advance of the event the sentence is about. On the tenseless theory of time, it is simply a mistake to suppose that just because a sentence is temporally qualified, i.e., is true at a certain time, or at every time, it follows that it corresponds to a fact that is a *temporal fact*, i.e., a fact that exists in time or at every time.

PHIL: So, as the detenser interprets it, the principle of the excluded middle does not imply that the future is "already laid up" in the present, and it does not imply that no person has free will.

IVAN: Exactly.

NOTES

1 Steven Cahn (1967) *Fate, Logic and Time*, New Haven, CT: Yale University Press, reprinted Atascadero, CA: Ridgeview Publishing Company, 1982.
2 Richard Taylor (1960) "I Can," *The Philosophical Review* 69: 78–89; the quoted passage occurs on p. 80.
3 Donald C. Williams (1951) "The Seafight Tomorrow," in P. Henle, H. M. Kallen and S. K. Langer (eds) *Structure, Meaning and Method: Essays in Honor of Henry M. Sheffer*, New York: The Liberal Arts Press, pp. 282, 306.
4 Aristotle (1941) *De Interpretatione*, in *The Basic Works of Aristotle*, ed. Richard McKeon, New York: Random House.
5 Donald C. Williams, "The Seafight Tomorrow," p. 282.
6 J. S. Mackenzie (1912) "Eternity," in J. Hastings (ed.), *Encyclopedia of Religion and Ethics*, New York: Scribner, p. 404.

GLOSSARY OF TERMS

B-statements Statements whose truth or falsity is independent of the time at which they are uttered, thought or written.

Correspondence theory of temporal truth A proposition is true in virtue of corresponding to a fact that exists at the time at which the proposition is expressed.

Eternal$_1$ An entity that exists throughout all of time. This conception of eternal is implied by the "non-temporal duration" definition of eternity, but does not imply it because the duration of an eternal$_1$ entity may consist of stages that succeed one another.

Eternal$_2$ An entity that exists outside of time. Thus, an eternal$_2$ entity is neither past, present nor future, and neither earlier, later nor simultaneous with any other entity.

Eternal$_3$ An entity that does not exist in time, but time (i.e., temporal relations between events) exists in it.

Fatalism The view that the laws of logic entail that no person has free will, that there do not now exist alternative possibilities for the future.

Law of the excluded middle For any proposition P, P is either true, or if not true, then false.

STUDY QUESTIONS

1 What is the argument for fatalism?
2 What are the presuppositions of the fatalist argument?
3 How can one argue that the tenseless theory of time entails fatalism?
4 How are Ivan's three definitions of "eternal" in this dialogue related to his definitions of "eternity" in Dialogue 4?
5 How can the detenser be defended against the charge of fatalism? Evaluate that defense.
6 How might a tenser avoid fatalism?

FURTHER READING

Bernstein, Mark H. (1992) *Fatalism*, Lincoln: University of Nebraska Press.
A good discussion of the role of time, truth and freedom in the problem of fatalism.
Cahn, Steven (1967) *Fate, Logic and Time*, New Haven, CT: Yale University Press, reprinted Atascadero, CA: Ridgeview Publishing Company, 1982.
Argues that the fatalism is unavoidable on the tenseless theory and so adopts a tensed theory of time and truth to avoid it.
MacKenzie, J. S. (1912) "Eternity," in J. Hastings (ed.) *Encyclopedia of Religion and Ethics*, New York, Scribner.
A proponent of the view that time is an eternal whole whose parts are in time.
Taylor, Richard (1991) *Metaphysics*, Englewood Cliffs, NJ: Prentice Hall.
Argues for the fatalist conclusion.
White, Michael (1987) *Agency and Integrality: Philosophical Themes in the Ancient Discussions of Determinism and Responsibility*, Dordrecht: Reidel.
A historically based treatment of fatalism.
Williams, Donald C. (1951) "The Sea-Fight Tomorrow," in Paul Henle, H. M. Kallen and S. K. Langer (eds) *Structure, Meaning and Method: Essays in Honor of Henry M. Sheffer*, New York: The Liberal Arts Press.
A classic defense of the tenseless theory of time against Aristotle's argument for fatalism.

Dialogue 10

God, time and freedom

SOPHIA: Several of the topics we have discussed over the past few days are connected to an issue that has puzzled me quite a bit recently.

IVAN: What issue is that?

SOPHIA: It is really a whole host of issues which can be conveniently summarized under the rubric, the dilemma of human freedom and God's knowledge of the future (called "divine foreknowledge").

PHIL: Could you explain the dilemma to us?

SOPHIA: Yes, but before I do I want to explain why it is such an important topic for those who have ever pondered the question, "Does God exist?"

PHIL: Go ahead.

SOPHIA: Philosophers have wondered how divine foreknowledge and human freedom could be compatible, and if they are not compatible the existence of God is in serious jeopardy.

PHIL: Why would the incompatibility of foreknowledge and freedom cast doubt on the existence of God?

SOPHIA: For one thing, God is defined, in part, as omniscient and therefore, whatever there is to know, God knows it. Thus, if it turns out that there are truths about reality, such as, "I will play with my children tonight" that God did not know, then God would not be God. After all, God knows all, so he must know what I will do and when I will do it. He must know not only what I will do tonight, but what I will do tomorrow, and every day, hour and minute thereafter. If God didn't have such knowledge then God would not be omniscient, and if God lacked omniscience then he (or she) would not exist.

PHIL: Agreed. God is essentially omniscient.

SOPHIA: It is also evident that if God exists, then humans must possess freedom of the will. The reason for this stems directly from the need for humanly free agents in virtually all solutions to the problem of evil.

PHIL: Could you quickly run through the problem of evil and explain why free will is necessary for its resolution?

SOPHIA: According to the problem of evil, at least in its simplest form, the existence of God and the existence of evil are incompatible. For if an all-perfect God created the universe, then presumably his creation would be perfect. For how could a perfect artisan create something that was imperfect? Any imperfection in the creation would reflect some defect in its creator, but God is without defect. Therefore, it would seem to follow that if God exists, and created the universe, then the universe must be perfect. And yet, God's creation (the universe) is not perfect. Leaving aside natural evil, or the human suffering that is the result of natural disasters such as earthquakes, famine, tornados, floods and the like, it is clear that there is moral evil. Moral evil is the pain and suffering that results from humans inflicting evil upon one another. Given that the universe is not perfect, it seems to follow that God is not the all-perfect creator of the universe. Since, however, God is defined as the all-perfect creator of the universe, we appear to be forced to the conclusion that God does not exist.

PHIL: Aren't there many ways out of this argument?

SOPHIA: Yes, but the most prevalent is the free will solution. According to it, the evil found in the universe is not the result of God. Rather, it is due to the misuse of a faculty that God has given us, namely, the faculty of *free will*. Thus, human freedom is essential for those who try to make sense of evil in a universe created by an all-good God. So, if God is to exist as an all-knowing, all-good being, the universe must contain humanly free agents, and God must have foreknowledge of how humanly free agents will act. In short, both divine foreknowledge and human freedom are necessary for the existence of God.

IVAN: I see, but tell me what difficulties arise when we begin to reflect on the conceptions of foreknowledge and freedom.

SOPHIA: One is that divine foreknowledge and human freedom are incompatible. In other words, if God knows what we are going to do before we do it, then our actions are not free. But if our actions are free, then God could not have known them before they occur.

PHIL: You are going to have to explain why there is an incompatibility because I don't see it. I know you will listen to what I have to say until I am finished talking, but it hardly follows from my knowing what you will do, that your listening to me is not free. To know that something will take place is not the same thing as determining or causing it to happen. Why, then, are you worried about the compatibility of God's foreknowledge and human free will?

SOPHIA: In part because of the great difference that exists between God's knowledge of what I will do and your knowledge of what I will do. You *can*, but God *cannot* be mistaken. For all you *know* I could decide not to continue listening to you. At any time I could walk out of the room while you are talking, and if I do then you did not really have knowledge of my future action at all. I do not doubt that your very rational and highly probable (but possibly mistaken) belief that I will continue to listen is compatible with my action being free. What I do doubt is whether or not God's *infallible*, unmistakably true belief of what I will do in the future is compatible with my future actions being free.

IVAN: Why does that puzzle you? Why do you think God's infallible knowledge would render human freedom illusory?

SOPHIA: In order to answer this question we need to define some terms. Let's begin with the notion of "freedom." There are two different conceptions of human freedom that we need to distinguish. According to the first, an individual freely chooses to perform an action A when the individual wants to do action A, and there is nothing that prevents her from doing A. On this conception, to act freely involves nothing more and nothing less than acting in accordance with what one wishes, wants or chooses to do. Thus, if I want to leave the room, but am compelled to remain (because I am suddenly tied up!), then my staying in the room is not a free action, but if I want to leave the room and nobody prevents me from doing so, then my action is free.

IVAN: That is one notion of freedom, but it differs from the one we considered when we were discussing the topic of fatalism. If you remember, yesterday we were assuming that a human action is free only if at the time just before we do it, it is in our power to do otherwise. For example, I am now talking with you of my own free will because I choose to talk to you. But it was in my power not to have this conversation, since a moment ago I could have decided to do something else, say, go out for a jog. Indeed, at any time during our conversation it is within my power to stop talking, and for that reason my talking to you now may be said to be done freely.

SOPHIA: This seems to be a somewhat stronger notion of freedom, since I may be doing what I want and yet not have the power to do otherwise. For example, I may have chosen to stay here and talk to you, but unbeknownst to me the door to the room is bolted shut, so I could not leave the room even if I wanted. In this case, since it is physically impossible for me to leave the room, we might plausibly maintain my remaining is not a free action, even though I am doing what I want.

IVAN: Suppose we assume that the power to do otherwise is necessary for human freedom. Why is that conception of freedom incompatible with divine foreknowledge?

SOPHIA: To see why we need to consider another definition. Suppose we define God's omniscience, his all-knowingness, in the following way: for any proposition *P*, God knows whether *P* is true or false. Thus, for any proposition about the past (e.g., I awoke at 7:30 a.m., August 1, 1993), the present (I am talking to Ivan at 2:00 p.m., August 1, 1994) or the future (e.g., I will give an exam in Phl. 101 on October 3, 1995), God knows whether it is true or false.

IVAN: That seems plausible enough. If God really knows everything, then he must know what *has* happened, what *is* happening and what *will* happen. To limit God's knowledge in any way would be an imperfection, but God is an all-perfect being. Thus, if God is omniscient, then God knows the truth or falsity of every proposition.

SOPHIA: Finally, we need an intuitively obvious assumption about the past, namely, that the past is not in our power to change. Thus, if something has already happened then humanly free agents such as you and I do not have it within our power to undo it. For example, having awoken at 7:30 a.m., it is no longer possible that I could have I awoken at 7:00 a.m.. This assumption is expressed colloquially by the expression, "Don't cry over spilt milk." Since the past is closed or fixed, we can regret our past actions and feel pride over our past accomplishments, but we cannot change them.

STEVE: Yes, the past is closed but the future is open. We can change it, or rather we can make it what we want it to be by our free choices. Admittedly, once we actually make a choice, some of our possibilities for the future are gone, but the choice of what possibilities to actualize is, to a great extent, up to us.

SOPHIA: It does seem reasonable to think that way, and I can't help believing that it is true. Nevertheless, on the basis of the definitions and assumption just mentioned, there is reason to believe that we do not have free will.

IVAN: What is the argument?

SOPHIA: I want to get it right, so I hope you do not mind if I explain it in a series of steps.

IVAN: Go right ahead. Perhaps you could put it on the blackboard.

SOPHIA: OK. Here it is:

1 Since God is omniscient, he now knows, at 10:00 a.m., January 27, 1993 (t_1) that I will stop talking at 11:00 a.m. January 28, 1993 (t_3). In other words, *before* I complete this conversation, it is a *fact* that God *knows* that I will stop talking at t_3.
2 If, however, I am really free at t_2 (that is, if I could do something other than stop talking at t_3), then it is within my power to bring it about that what God knows is false, or it is within my power to change the past, by annulling a fact about the past, namely, that at t_1 God knew that I would stop talking at t_3.
3 But neither of those possibilities is acceptable. Since God is omniscient he cannot make mistakes (or hold false beliefs) about what he foresees in the future. And since the past is fixed, it is not within my power to do other than what, as a matter of past fact, God knew I would do.
4 Thus, on the reasonable assumption that God's omniscience implies foreknowledge of human actions, it follows that neither I, nor any other humanly "free" agent, has it within their power to do anything other than what God knows we will do.

In short, if God knows *before* I perform a certain action what action I will perform, then it is not in my power to perform any other action than the one he knows. For I can neither render God's knowledge, his true justified beliefs, erroneous, nor can

I, at t_3, change what he knows at t_1, since to do so would be to change the past and that is impossible. Thus, either human freedom is an illusion (or we must understand it differently), or God is not omniscient (or we must understand it differently).

IVAN: So what you are saying amounts to this: if at a certain past time (t_1) God truly believes that I would do X at t_3, then I do not have it within my power to do otherwise. For, at t_2 it is already a fact that God truly believes that I will do X at t_3. It became a fact (at t_1) when he believed it, and once something is a fact it is always a fact. This follows from the assumption that the past is closed, fixed and unalterable. If, however, neither I nor anyone else has the power to change facts about the past, I cannot do other than what God knew that I would do. Given that the ability to do otherwise is a condition of freedom, it follows from this argument that I am not free.

SOPHIA: Yes, that is one way of formulating the difficulty. Shall we consider some attempts to resolve it?

IVAN: Yes.

SOPHIA: Boethius, Aquinas and others have claimed that God is eternal, meaning that he exists outside of time. On their view God does not, strictly speaking, have *fore*knowledge. Foreknowledge presupposes that God knows what we will do *before* we do it, but if God is outside of time then God sees and knows everything timelessly. We could say that he sees everything as we see the present, but from the fact that I am now seeing you listen to me it does not follow that you have not freely chosen to listen. In other words, since my perception of you as listening is compatible with your previously having the power to not listen, my present perception of you listening is compatible with human freedom. And since God sees all of history as present, God's seeing what I will do is compatible with my freely having chosen to do it.

PHIL: Let me see if I get it. Boethius argued, in effect, that divine *foreknowledge* is a misnomer. God is all-knowing, but he does not have foreknowledge because he is *not in time*. God's eternity is not an everlasting eternity (always has and always will exist), but God's eternity is *timeless, outside of time altogether*. Thomas Aquinas, who thought God was eternal, expressed this view by saying: "He who goes along the road does not see those who come after him; whereas he who sees the whole road from a height sees at once all those traveling on it."[1] In other words, what is for us the future, and known only as we make our way towards it, is for God a part of the entire temporal history of the universe which he sees all at once.

Viewing God's relation to time, and the sequence of events in time in this way provides a solution to the problem of divine foreknowledge and human freedom. For the fact that some action is known in the present, when it is occurring, has no tendency to show that it was not done freely. If I now see a student walk into the classroom, that in no way implies that she was not free to do otherwise the moment before she entered. Similarly, God's seeing what I will do in the future is analogous to my seeing what you are doing in the present, and so does not curtail freedom of action.

SOPHIA: That's the view exactly.

IVAN: Boethius's solution raises many interesting questions, the most pressing of which is whether or not a consistent interpretation of God's *timeless* eternity can be given. For clearly, if God does not exist outside of time, then this particular solution will not work. Unfortunately, our discussion of a few days ago cast considerable doubt on the possibility of understanding God's eternity to be timeless.

PHIL: I also wonder if this conception of God's eternity can be reconciled with the tensed theory of time, according to which those events which are NOW have a special ontological status. After all, if God sees all events at once, how can the ontological distinctions between past, present and future events be preserved? Furthermore, if God sees the history of the universe in one timeless act, how can he know which ones are happening *now*, and without such knowledge how can God be omniscient?

SOPHIA: The questions you raise are certainly worthy of discussion, and they bring to mind another problem with Boethius's view.

ALICE: What problem do you have in mind?

SOPHIA: Clearly, Boethius's solution rests on an analogy between what is presented or seen by humans and what is seen by God, but the analogy is faulty. What is present*ed* to humans is *temporally present*; it is (roughly) *simultaneous* with the act of seeing. However, what is presented in God's timeless act of seeing is not temporally present. Thus, there is a fundamental disanalogy between what God perceives as a (never-changing) present, and what humans perceive as an (always-changing) present.

What's worse, suppose the analogy were correct, so that what God perceives is temporally present. In that case, since God perceives all of time, it would follow that all of time, that is, all of the events in the history of the universe, would exist NOW and that is absurd.

PHIL: If your reasoning is sound then the Boethian view does seem to be faced with a dilemma which could be put this way: either God's perception of total world history is analogous to our perception of presently existing events or it is not. If it is not, then it is unclear why God's (timeless) perception of human actions should be thought to solve the foreknowledge dilemma. On the other hand, if it is analogous then it follows that what is presented to God is either simultaneous with his act of presentation or possesses the property of *being present* (or both). And that is absurd, since it implies that the entire course of history is occurring NOW. Perhaps there is a plausible interpretation of the Boethian view, but as it stands I find it unacceptable.

IVAN: If the Boethian solution is inadequate, what is to be done? Maybe there is some way to preserve the compatibility of God's omniscience and human freedom if we maintain that God's eternity is a temporal eternity.

SOPHIA: There is one solution that is a bit radical, but worth mentioning. It assumes God's eternity is temporal, and it brings into focus an alleged connection between the foreknowledge issue and the nature of time.

IVAN: What solution is that?

SOPHIA: Before I spell it out, let me put the dilemma of freedom and foreknowledge in a slightly different light so as to bring out more clearly its connection to issues in time.

IVAN: Fine, go ahead.

SOPHIA: Suppose we conceive of God's foreknowledge as involving a telescope that allows God to peer into the future and observe what we will do before we do it. In other words, just as God can see what has occurred, and what is now occurring, he can see what will occur. It has appeared to some that if God had foreknowledge in this sense, then the future, like the past, would *fully exist*, and thus be closed, fixed and unalterable. In that case, however, human freedom would be lost. On the other hand, if we are free then arguably the future must be an open realm of possibilities that have no actual reality until the time some of them come into existence.

IVAN: I think I see your point. Suppose we look at it this way. Human freedom and responsibility require there be a fundamental difference between the past and the future: the future must be open and the past closed. According to J. R. Lucas, for example, only the tensed view of time which treats the past as fixed and the future as open

> allows for freedom and responsibility and creativity, and since they [our past deeds] cannot be undone or conjured out of existence, it acknowledges the everlasting significance of our deeds . . . We are by our decisions in the face of other men's actions and chance circumstances weaving the web of history on the loom of natural necessity. What is already woven is part of the fabric of the universe, but what is still unwoven has as yet no substantial reality.[2]

SOPHIA: That sounds right. If we are assuming that freedom demands an open future, the dilemma of reconciling divine foreknowledge and human freedom may be stated this way. How can God have foreknowledge of an open future (that is, of what as yet does not belong to reality), if, at present, the future is nothing more than a set of possibilities, "nothing substantial," but only a "species of unreality." Put differently, if the future does not yet exist, and the telescope into the future through which God looks can yield nothing definite, but only an indefinite number of possible futures, how can God know what I or anyone else will *actually* do? And without such knowledge how can God be the all-knowing being he or she is supposed to be? On the other hand, if the future does exist, God can know it; but then how can we be free to create it?

ALICE: I think I see the problem now. Freedom requires that the future does not (yet) belong to the sum total of reality, whereas divine foreknowledge requires that the future does already belong to the sum total of reality. Thus, freedom excludes foreknowledge and foreknowledge excludes freedom.

SOPHIA: Yes, that's it. There have been a variety of solutions that have been offered to this version of foreknowledge dilemma, but let me just mention two.

IVAN: What are they?

SOPHIA: One interesting view, a secular version of which was championed by Aristotle, shares with the Boethian solution the idea that God does not have foreknowledge. On this view, God does not have foreknowledge because the future does not exist and so there is nothing about the future for God to know. God exists in time, and knows the truth or falsity of all propositions about the present and the past. However, propositions about the future, such as "I will go to the movies tomorrow," are neither true nor false (and so are not known by God), because there is nothing in reality for them to correspond to.

IVAN: Doesn't this view come in conflict with the belief in God's omniscience?

SOPHIA: Not really. We defined God's omniscience in terms of knowing all true propositions. This view does not conflict with our definition because propositions about the future are not true. God does know all true propositions, he just does not know what will happen in the future. There is nothing to be known about the future. Thus, we have preserved the open future, an alleged requirement for human freedom, and God's omniscience.

IVAN: I suppose that this is a possible view to take, and it would remove the dilemma, but it seems rather radical. After all, if God planned the universe then presumably he would know what would happen. Moreover, it seems intuitively true that every proposition is either true, or if not true, then false. The view you are suggesting must deny this principle.

SOPHIA: Yes it must, and for that reason it may be thought objectionable.

ALICE: Are there any other solutions to the foreknowledge dilemma?

SOPHIA: Yes, another is associated with the fourteenth-century philosopher, William Ockham. He maintained that God can know that a proposition about the future is true, even though the state of affairs in virtue of which it is true is not yet settled or actual. If this view can be rendered plausible, we could preserve the openness of the future, human freedom, and divine foreknowledge. For the will is free so long as I can do otherwise, and I can do otherwise so long as what I will do is not settled before I act.

ALICE: I must admit, this view is attractive. If God can know today that, for example, I am going to the movies tomorrow, even though the future is now bereft of content, the foreknowledge and freedom dilemma seems readily solvable. What I find puzzling is how God can know, at t_1, that "I will go to the movies tomorrow (at t_3)" *is true* at t_1. There is no tenseless state of affairs such as *I go to the movies later than* t_1 or *I am at the movies at* t_3. Nor does there exist, at t_1, the tensed state of affairs *I will go to the movies*. Thus, if the future does not exist and the tenseless theory of time is false, there simply is no basis for God's knowing, at t_1, that a proposition about the future is true. That is, he does not know whether "I will go to the movies tomorrow (at t_3)" or "I will not go to the movies tomorrow (at t_3)" is true *today*. Without such knowledge, however, the Ockhamist attempt to reconcile foreknowledge with the open future conception of freedom is unsuccessful.

IVAN: I agree there is a serious problem here if one denies reality to the future. On the other hand, it seems to me that there is a relatively easy solution to the problem you have been raising if we adopt the tenseless theory of time.

SOPHIA: What is your solution?

IVAN: In order to explain it, I need to clear away some errors that have infested the discussion. So far we have assumed that human freedom requires an "open future," which we interpreted to mean that the present and the past exist, but the future does not exist.

SOPHIA: Yes, we have made that assumption.

IVAN: I don't think that assumption is a reasonable one to make, since I don't think it is required for our human freedom. Let me explain. When we say that the past is closed and the future is open, I suggest that what we mean or should mean is that the past "already exists," but the future does not already exist. The past "already exists" because the events of which it is composed occur *before* now. Of course, if one believes the future is as closed as the past and consequently, that it "already exists," then one is likely to assume that future events exist *before* the times at which they occur. Clearly, if the future "already exists" in that sense then our entire lives are like a film in which events are laid out beforehand. All our deeds and decisions are fixed in advance and we are only puppets in the universal drama. However, neither the tenseless theory of time nor causal determinism is incompatible with the view that the future is open, i.e., with the view that the future does not *already* exist. Thus, we need not deny the reality of the future to preserve freedom.

ALICE: Could you explain why neither the tenseless view nor determinism imply that the future is closed?

IVAN: Some of my argument goes over material covered in our discussion of fatalism, but it is important to restate it here. On the tenseless view, to say, at time t, that the future is real, is to say that "There are events later than time t." Of course it does not follow that events later than t *already exist* at t; nor do the events that occur at times later than t somehow "pre-exist" at t. Thus, the tenseless view does not imply that the future already exists, i.e., that future events exist before they take place.

Furthermore, the tenseless view does not imply that future events are causally determined but only that they are determinate, having (tenselessly) the properties that they do at a certain clock time.

PHIL: Could you more fully explain the distinction between causally determined and determinate?

IVAN: Certainly. To say that a future event is causally determined is to say that it is correlated with some past event or events by laws of regular sequence. Thus, for example, as I see a rock flying through the air toward a mirror, the future event of the mirror breaking is causally determined. More generally, if the future is determined then there is a sufficient explanation of future events in terms of past events and laws of nature. To say that the future is determinate is a much weaker thesis. The determinateness of the future implies only that propositions about the future are either true or false. If the future is determinate then future events exist with the properties that they have at the time and place at which they have them, but that does not imply determinism. Events cannot be determined without being determinate, but they can be determinate without being determined. For example, it may be the case that I will eat breakfast at 6:00 a.m. on October 23, 1995, and yet it not be the case that there are prior events that causally determine me to do so.

Moreover, even if the tenseless view did entail determinism, which it does not, it would not follow that such a view entails that the future history of the universe pre-exists in the present, since determinism does not imply that the future is already laid up in the present. That a future state or event is uniquely specified by a present state (or set of events) does not detract from the future state being later than the present state. In other words, the existence of causally sufficient conditions for the future does not imply that the future is present. Thus, in so far as the openness of the future means that the future does not already exist in the present, neither the tenseless theory nor determinism involves a denial of the openness of the future.

ALICE: What does all this have to do with a resolution of the foreknowledge and freedom dilemma?

IVAN: I see a way of maintaining that (1) the future is open, (2) causal determinism is the case, (3) the tenseless theory is true, (4) humans have free will and (5) God has foreknowledge! Clearly if each of these propositions is true, we have reconciled divine foreknowledge with human freedom.

SOPHIA: Agreed, but can they be reconciled?

IVAN: I believe so. Will you agree that if the tenseless theory is true, then God can have foreknowledge?

SOPHIA: Certainly. If all past, present and future events are equally real, then God can at any time see any and all of them.

IVAN: Then the solution to the foreknowledge dilemma is simple. Suppose God knows at t, that I will (at t_3) make a conscious choice, based on my wants and desires. For example, God knows that at 2:00 p.m. on August 30, 1993, I want to go to the movies at 9:00 p.m. on August 30, 1993, and he knows that I (freely) choose to go to the movies at 9:00 p.m. on August 30, 1993. In knowing those two facts he knows that I will freely choose to go to the movies. Since God knows what I will freely choose to do, and his knowledge is infallible, it must be the case that I freely choose to do it.

SOPHIA: You are erroneously assuming that an action is free if you are doing what you want. For if your going to the movies is determined by prior circumstances then your choice was not free. Don't you know that free will and determinism are incompatible? Of course, if your conception of freedom is right, then your solution is viable. But I reject your conception of freedom.

IVAN: So the crucial question concerns whose conception of freedom is the correct one. Shall we proceed to discuss that issue?

SOPHIA: Let's. To do so will allow us to connect the free will–determinism issue with another we considered recently, namely, the nature of the self.

NOTES

1 St Thomas Aquinas (1952) *The Summa Theologiae*, trans. by the Fathers of the English Dominican, Chicago: *Encyclopaedia Britannica*, Ia, q. 14, A. 13, reply obj. 3, p. 88.
2 J. R. Lucas (1989) *The Future: An Essay on God, Temporality and Truth*, New York: Blackwell, pp. 8–9.

GLOSSARY OF TERMS

Causal determinism For every event that occurs there is some condition (event) or set of conditions (events) sufficient to bring about that event.
Closed past Once an event has happened, it is fixed, necessary and cannot be otherwise.
Determinate An object is determinate if its properties are completely specifiable at each moment of its existence.
Foreknowledge Knowledge of an event at a time before the time at which the event occurs.
Free will solution An attempt to solve the problem of evil by maintaining that the evil that exists in the universe is not due to God, but to the choices of humanly free agents.
Open future The future does not already exist.
Problem of evil How can an all-good, all-knowing, all-powerful God create a universe that contains so much evil in it?
Two conceptions of freedom (1) An action is free if it is what we want to do. (2) An action is free only if it could have been otherwise.

STUDY QUESTIONS

1 What are two different formulations of the dilemma of freedom and foreknowledge?
2 What is the relation between the problem of evil and the dilemma of freedom and foreknowledge?
3 What is the Boethian solution to the problem of divine foreknowledge and how might it be criticized?

4 What are two other solutions to the foreknowledge dilemma? Do you find either acceptable?
5 In what ways, if any, do the different solutions to the foreknowledge dilemma depend on one's views on time and freedom?

FURTHER READING

*Craig, William Lane (1991) *Divine Foreknowledge and Human Freedom*, Leiden: E. J. Brill.

Offers a solution to the foreknowledge dilemma within the context of a tensed theory of time.

Fischer, John Martin (ed.) (1989) *God, Foreknowledge, and Freedom*, Stanford, CA: Stanford University Press.

A collection of articles on the issues raised in this dialogue.

Hasker, William (1989) *God, Time, and Knowledge*, Ithaca, NY: Cornell University Press.

Avoids the problem of freedom and foreknowledge by restricting the sense in which God's omniscience applies to the future.

*Lucas, J. R. (1989) *The Future: An Essay on God, Temporality, and Truth*, New York: Blackwell.

A defense of the "open future" version of the tensed theory according to which the past and present are real, but the future contains nothing but possiblities.

Mellor, D. H. (1986) "History Without the Flow of Time," *Neue Zeitschift für systematische Theologie und Religionsphilosophie* 28: 68–76.

Explains how the foreknowledge dilemma can be resolved on the tenseless theory.

Rowe, William L. (1993) *Philosophy of Religion: An Introduction* (2nd edn) Belmont, CA: Wadsworth Publishing Company, pp. 154–69.

Contains a useful discussion of the conception of freedom upon which the problem of divine foreknowledge is based, and discusses some of the main solutions to it.

*Zagzebski, Linda Trinkhaus (1991) *The Dilemma of Freedom and Foreknowledge*, New York: Oxford University Press.

This book discusses the major solutions offered to the foreknowledge dilemma, proposes three possible solutions of its own, and raises a new foreknowledge dilemma.

Dialogue 11

Freedom, determinism and responsibility

SOPHIA: So far we have presented two different kinds of challenge to the reality of freedom. The first is based on logic and the law of excluded middle, and the second is based on religion and the nature of divine foreknowledge. There is, however, a third challenge to freedom that is equally interesting.

IVAN: What challenge do you have in mind?

SOPHIA: It is a challenge that comes not from logic or religion, but from nature, or more specifically, from the connections most people believe exist among events in nature. Most people believe events have causes, and the causes of events "necessitate" the occurrence of those events. That is, given certain laws and initial conditions, then what happens is determined.

ALICE: Could you give me an example of laws and initial conditions determining something?

SOPHIA: Well, it is a law that if under laboratory conditions water is cooled to a temperature of or below 32 degrees Fahrenheit then water freezes. Thus, if the conditions are such that the environment of a cup of water is below 32 degrees then its freezing is determined.

Now, although science has not discovered all the laws governing human actions, it would appear human behavior, like non-human behavior, is causally determined. In other words, we believe that people's character and motives, together with the circumstances in which they are placed, will cause them to behave in a certain way.

PHIL: That might be true in general, but the most basic laws of modern physics are not all deterministic. It is a feature of quantum mechanics that, in the micro-world, some events are random (or certain initial conditions do not necessitate specific later states). For example, the behavior of an isolated micro-particle cannot always be predicted with certainty because its position and velocity are not both determinate.

SOPHIA: True, but these quantum indeterminacies do not usually affect the behavior of large collections of particles like human bodies and brains. Human bodies and brains could be causally deterministic, even if events in the micro-world are random. If, then, human actions are causally determined, it might seem, contrary to common beliefs, that people are not free.

IVAN: The issue of whether our intuitive belief that we are free conflicts with our belief in determinism connects with the question we raised at the end of our last discussion. Remember, we were then wondering whether freedom should be understood solely in terms of doing what one wants or wishes to do or whether, in addition, a truly free act must be one that could have been otherwise. It should not surprise us if the question of the compatibility or incompatibility of causal determinism and human freedom hinges largely on which conception of freedom is correct. For example, those who maintain that the ability to do otherwise is a necessary condition of freedom also typically believe that free will and determinism are incompatible. Others, like Hume, thought that the entire free will versus determinism debate was a verbal dispute that could be resolved or dissolved once we properly defined our terms. That is, Hume thought that if we define "freedom" in terms of acting in accordance with one's will, then freedom and determinism are compatible.

SOPHIA: Metaphysical disputes about freedom have important practical and ethical implications as well. We often praise or blame and reward or punish people for their actions. Philosophical theories about what makes it right to do this involve the notion of responsibility, and the idea of responsibility is usually tied to a person's being free. For example, punishment in the form of retribution is clearly wrong if we are not free but are compelled or forced to act as we do.

IVAN: There is an interesting legal case that can serve as a starting point for a discussion of determinism, freedom and responsibility. The case involves two young students (Nathan Leopold Jr and Richard Loeb) at the University of Michigan who, in 1924, unsuccessfully attempted to commit the "perfect crime." They did, in fact, kidnap and murder a 15-year-old boy, but were caught. They both came from wealthy families who retained the services of the most famous lawyer of the day, Clarence Darrow. Since there was no question but that Leopold and Loeb committed the crime, Darrow's plea was that although guilty as charged, they should be spared their lives.

ALICE: Given the different attitudes concerning the death penalty in the 1920s, and given the hideous nature of their crime, how did Darrow argue his case?

IVAN: He tried to convince the jury, through a consideration of the childhood, boyhood and youth of Leopold and Loeb that their behavior was the inevitable result of forces over which they had no control. In other words, he argued, in effect, that the environment, upbringing and heredity of the two boys, together with the circumstances in which they found themselves, *determined* their actions, the implication being that if their actions were determined, then they were unfree. Since freedom is a precondition of responsibility, Darrow concluded that the boys should not be held responsible for their actions.

In one part of his defense Darrow said:

> Nature is strong and she is pitiless. She works in her own mysterious way, and we are her victims. We have not much to do with it ourselves . . . What had this boy to do with it? He was not his own father; he was not his own mother; he was not his own grandparents. All of this was handed to him . . . If there is responsibility anywhere, it is back of him; somewhere in the infinite number of his ancestors, or in his surroundings, or in both. And I submit, Your Honor, that under every principle of natural justice, under every principle of conscience, of right, and of law, he should not be made responsible for the acts of someone else . . .[1]

PHIL: That is the most ridiculous argument I ever heard in my life. It's just a cop-out, a way of avoiding responsibility. Anybody can literally get away with murder by committing a crime and saying that they were caused to do it by events in their past.

IVAN: It does seem to be rather outlandish, but Darrow did succeed in keeping his clients out of the electric chair. I raise this case, not because I am sympathetic with the argument, but because it brings to light some of the crucial issues surrounding the topics we are considering.

ALICE: What are they?

IVAN: Darrow's argument reflects a position known as *hard determinism*. According to this position, if our actions are causally determined by antecedent events, then they are not free. Since, however, hard determinists believe our actions are determined, they conclude that human freedom is an illusion and consequently, we are not responsible for our actions.

PHIL: But why do they believe that if an action is causally determined, then it is not free?

IVAN: This is where the incompatibilist definition of freedom comes in. Hard determinists, like incompatibilists in general, maintain that the power to do otherwise is a necessary condition of freedom. They also believe that if an action is caused by prior events then, given those events, it could not have been otherwise. Thus, they conclude that freedom and determinism are incompatible.

ALICE: Let me make sure I understand the line of reasoning so far. The hard determinist maintains that an action A is a free action only if instead of action A, action B could have occurred.

IVAN: Yes.

ALICE: It is then argued that if action A is determined or caused or necessitated by prior circumstances then it couldn't be otherwise. The idea is that if an action is *caused* then given the cause the action *must* follow; there could not have been any other action. However, if an action could not be otherwise, then it is not free. And, of course, if an action is not free then we cannot be held responsible for it.

IVAN: Yes, that's the logic, but we can get at the intuitive idea behind their position in a slightly different way. We ordinarily believe that an action is performed freely only if it originates with us. Aristotle once said that an action is voluntary when the moving principle is in the agent and so it is in the power of the person himself or herself to do or not to do the action. If however, our actions are caused by our constitution (e.g., our character and motives) and the external circumstances in which we find ourselves, and our being in those circumstances and having that constitution can be traced back to causes that lie outside us, e.g., our parents, grandparents, teachers, genetic make-up and the like, then ultimately, the cause of our action does not lie within us (the moving principle is not in the agent). But then the person does not have it within his or her power to do other than what in fact he or she does do, and so is unfree.

ALICE: So, since Darrow accepts determinism, he denies both human freedom and responsibility.

IVAN: Yes, but I disagree.

ALICE: Why?

IVAN: I believe that determinism is a frame of reference for understanding the world and that determinism can be reconciled with human freedom.

PHIL: What do you mean by saying that "determinism is a frame of reference for understanding the world"?

IVAN: Well, we do believe, concerning things that we experience in nature, that there is a cause for why they are so and not otherwise. For example, if I see water boiling I believe that its being heated caused it to boil. Of course there are many cases when we do not *know* the cause. For example, we do not know what caused the AIDS virus to first appear, but we do believe that there was a cause, and scientists are hard at work to find it out. Furthermore, there is every reason to believe that the same is true of human actions. Of course, you might think you know what a person will do in a given situation and that person may surprise you. But even when that happens we still believe that there is some reason, in the sense of cause, for why the person acts differently. We don't believe that such unusual behavior has no cause, a complete accident without rhyme or reason.

ALICE: Given that determinism is a frame of reference that most of us implicitly assume in our daily dealings with the world and people around us, why are so many philosophers convinced that determinism is incompatible with freedom?

IVAN: In response to your first question I suggest that the very word "determined," and the claim that if determinism is true then every event is "necessary," gives rise to the impression of its incompatibility with human freedom. For the words "necessarily determined" suggest that the cause *compels* or *forces* the effect to occur whether one wants it to occur or not. But I think that these suggestions are misleading. If we base our understanding of causation and necessity on our experience, then we see, as Hume argued, that there is no compulsion involved. Hume said that "A causes B," or that "A determines B," are statements whose truth involves nothing more than the constant conjunction of events of a certain type A being followed by events of type B. Upon having observed many instances of similar causes being followed by similar effects, (e.g., seeing a baby first fall and then cry) we conclude that falling caused this baby to cry. A further basis for the idea of necessity comes from the repeated experience of a constant conjunction which gives rise to a transition in the mind from the cause to the effect (when I see a baby fall I immediately think that it will cry). We should not conclude, however, that there is a necessary connection in nature, or a power in the cause that forces the effect to occur. If we think of causality as just describing lawful relations among the properties of individual things (e.g., whenever water has the property of being heated to 212 degrees then it has the property of boiling), then there should be no temptation to think of causation as compulsion, i.e., the picture that if our choices are caused then we are somehow overwhelmed by the past.

SOPHIA: It would take us too far afield to get into an extended discussion of causation, but I must point out that the Humean view of causation is questionable. For one thing, as it stands, it fails to distinguish between accidental generalities (or constant conjunctions that do not reflect causal connections) and lawful connections. For example, although night always follows day, it is not the case that night and day are causally connected. Leaving that issue aside, how do you propose to render freedom and determinism compatible?

IVAN: I was getting to that. Let's look at an example. Consider my present action. At this moment I am sitting before you and talking about the free will–determinism issue. As I see it, I am now acting freely. Why? Because last night I thought about coming to the office this morning to talk philosophy, and on the basis of such deliberation I made the choice to be here. Thus my being in the office right now is a *conscious choice* or decision that I made. No one forced me to choose to come here, and obviously no one prevented me from coming to the office because here I am. In short, if my action results from a conscious choice that I make, then my action is free, whereas if it results from a conscious choice that someone else makes then it is unfree. Alternatively, we could say, following Hume, that if I do as I will, then my action is free, whereas if my action is compelled by some external circumstance, or I am constrained to perform an action against my will, then it is unfree. Liberty or freedom is not to be contrasted with necessity or determinism, but with compulsion or constraint. Thus, even if my action is causally determined it may still be free.

PHIL: I think I understand you: it is not *that* our actions are caused but rather it is *how* they are caused that is relevant to whether or not they are free. If they are caused by me, i.e., by a conscious choice that I make, then they are free, whereas if they are caused by you or something external to me, then they are unfree. Since a free action is one for which we can be held responsible, on your view, freedom is compatible with both determinism and responsibility.

IVAN: I would go further and say that not only are determinism and responsibility compatible, but that without determinism no person can be held responsible for his or her actions. For if determinism is false, and our actions are not caused by our character and motives together with the circumstances in which we find ourselves, then our actions just happen for no reason. They are, as it were, arbitrary and capricious. But then, they can hardly be considered free. Since, however, freedom is a condition of moral responsibility, such indeterministic acts are ones for which we cannot be held responsible.

SOPHIA: I do not find this argument all that convincing, and will explain why when I discuss my own libertarian view. But before we turn to that, let me say I am still not satisfied with your attempt to reconcile freedom and determinism.

IVAN: Why not? What is wrong with it?

SOPHIA: There are two problems. First of all, you put a lot of weight on the idea that we make choices, and our actions are free if they flow from a conscious choice. But I don't understand how determinism is compatible with there being genuine choices at all. If your "choice" is determined by antecedent events, then how can you say you really have a choice? I believe that we do make genuine choices and for that reason reject determinism.

IVAN: So your argument amounts to this: (1) if determinism is true then none of us make *genuine choices*. (2) Since, however, we do make choices, (3) determinism must be false.

SOPHIA: Exactly.

IVAN: Your argument is based upon a confusion of two different meanings of the word "choice" or the idea of a "genuine choice." By "choice" I mean a certain process or state of events which occurs in human beings from time to time. This is a process that involves making a decision based on prior deliberation. In that sense it is clear and unproblematic that we do make choices, genuine choices. Moreover, in this sense of the term, determinism is compatible with our making genuine choices, for there is nothing in the nature of choices which implies that they are or are not caused by prior events.

On the other hand, if by "genuine choice" you mean a choice that is not causally determined, then it is not clear that we do make genuine choices. That is, if you just assert that a necessary condition of an event being a choice is that it is not causally determined, then you are simply assuming that the existence of choices is incompatible with determinism and so are begging the question. Thus, given either understanding of "choice" your argument rests upon a doubtful premise.

SOPHIA: I think you are missing the point of my argument, as will become apparent by considering a second argument against your view.

IVAN: What is it?

SOPHIA: My argument goes like this. An action is free only if it could have been otherwise. But if determinism is true, then we could never do anything other than what we do do. Therefore, if we are free then determinism must be false. Look at it this way. There are really two components in an action. There is the determination of the will to do X, and there is the actual event of doing X. In order for an action to be free, it is not enough that the event of doing X is the result of doing what one wills. It must also be the case that the will is free. But when is a will free? We have freedom of the will only when it is in our power to will to do X or to will not to do X. However, if determinism is the case, then given our motives, character and circumstances, we could not will or choose anything other than what we do will or choose, and therefore, we could not do anything other than what we actually do. Thus, if determinism is true, we are neither free nor responsible for our actions. Since, however, we are free and are responsible for our actions, determinism must be false. I just don't see how freedom and determinism can be reconciled, and since I affirm freedom I am a *libertarian*.

IVAN: As I shall suggest later, I doubt that the ability to do otherwise is a necessary condition of freedom or responsibility, but I will let that pass for the time being. What I now wish to question is the thesis that determinism is incompatible with the idea that we could do otherwise, for I don't think that they are incompatible.

SOPHIA: Could you please explain your reasoning to me?

IVAN: There is an ambiguity in the word "could" when we say that if an individual has free will then he or she could have done otherwise. In one sense of the word "could," if determinism is the case, and I did action A, then I could not have done otherwise. In another sense of the word, however, even though I performed action A, I could have done otherwise.

SOPHIA: What are the two senses of "could"?

IVAN: Suppose we say of a cat that did not climb the tree that she *could* have climbed the tree, but that a dog who did not climb the tree that she *could not* have climbed it. The notion of "could" in this context refers to what is causally possible. Given the physical constitution of the dog, she could not climb the tree; it is causally impossible. Since the cat could climb the tree, it is causally possible for the cat. On one interpretation, then, to say of an event X (my walking to school this morning) which did not happen (because I drove), that it *could* have happened means that it is causally possible. On this understanding, only what is causally possible could have been otherwise.

In this sense of the word, to say that "An event X which did not happen, could have happened," means that "X would have happened, if so-and-so had chosen or willed it to happen." Thus, for example, although I didn't walk to school this morning, I could have walked to school this morning, meaning that I would have walked to school, if I had chosen to do so.

SOPHIA: What is the second sense of "could"?

IVAN: In a second sense, an event X which did not happen could have happened if and only if not all events are determined by a cause that precedes it. Thus, if X occurs, then to say that something else could have happened means that X was not caused to happened by prior events. Clearly, in this sense of "could," if determinism is the case, then we could not do otherwise.

SOPHIA: How does that distinction help avoid my problem with your account of freedom and responsibility?

IVAN: Well, you are claiming that determinism is incompatible with freedom and responsibility, because a condition of freedom, namely, the ability to do otherwise, is not fulfilled if determinism is the case. What I am maintaining is that there is a sense in which I could have done other than what I do even if determinism is the case. For even if action A is determined, I could have done otherwise in that I *would* have done otherwise, had I made a different choice. It was causally possible for me to act differently if I chose differently.

SOPHIA: The problem with your argument is that your interpretation of "could have done otherwise" is inadequate. You say that "She could have done otherwise" means that "She would have done otherwise, if she had so chosen." I agree that if she had chosen differently then she would have acted differently, but that is a moot point because she couldn't have chosen differently. Since our choices are causally determined, they couldn't be otherwise, and if they couldn't be otherwise it hardly makes a difference to our freedom to say that if our choices had been different we would have acted differently.

IVAN: But I deny your claim that if determinism is the case, we could not have chosen or acted differently. To say that "She could have done otherwise" means that "She would have done otherwise, if she had made a different choice," and that is compatible with the action that one in fact performs being caused.

SOPHIA: But could she have chosen to act differently? I don't think so, if determinism is the case.

IVAN: But I do. To say that a person could have chosen to act differently means that a person would have acted differently, if the deliberation and reflection that led to the choice was different. You are simply assuming that if our actions are caused, then they could not be otherwise, but that just reflects the confusion between the two senses of "could" that we have been discussing.

SOPHIA: I still don't see how a determined action could ever be other than what it is, if it is determined to be just that action.

IVAN: I can see that you don't agree with this analysis of "could have done otherwise," but perhaps we could move on since later I will suggest that the ability to do otherwise should not be a necessary condition of human freedom and responsibility anyway. So, rather than try to continue defending the compatibilist approach now, I want to know what you would say about the alleged dilemma of freedom and determinism that we have been considering.

ALICE: I am afraid that I have lost sight of why exactly there is a dilemma concerning these two notions. Could you explain it again?

SOPHIA: Certainly. As I understand the issue, there is a prima facie problem with maintaining that determinism is false, and it is precisely the same problem as that which arises if determinism is true. As you know, I believe that if various events, such as a person's wants, beliefs and desires are the cause of a person's action, then we cannot hold that person responsible for the action since the person did not bring the action about. Of course, if the person brought about his or her wants, desires and wishes, then he or she could be held responsible, but if determinism is the case then our wants, desires and beliefs are caused by other events and not by the agent. So, it seems that freedom and responsibility require indeterminism.

Unfortunately, it does not help to maintain that actions are not caused by people's character and circumstances since if actions are totally uncaused, happening out of the blue, then once again no individual person is the cause of them and so no individual can be held responsible. If my left arm suddenly jerks up in the air, without being caused to do so, *I* cannot be held responsible for that event. Thus, the dilemma is that if every event is caused by some other event, then we are not free and hence not responsible. And if human actions are not caused, then our freedom is beyond our control, utterly spontaneous and so once again we cannot be held responsible.

IVAN: As you know, I believe that determinism and responsibility are compatible, but I agree the problem you pose does arise for the indeterminist. How do you propose to resolve it?

SOPHIA: It seems to me that to solve the problem we must adopt a certain conception of personal identity, namely, the substantialist conception. Persons are substances, not necessarily immaterial Cartesian substances, but substances nonetheless. Since I am a substance, and not an event or a succession of events, it is within my power to bring about or cause an event or action to take place without myself being caused to do so. If we ourselves bring about an action without being determined to do so by any prior cause, then the action is free and one for which we are wholly responsible. Each of us, on occasion, is an uncaused cause. Sometimes when we act we cause certain events to happen, and nothing or no one causes us to cause those events to happen.

PHIL: Your view does have important implications for the nature of personal identity. If we are a succession of events somehow related, as the relational view maintains, and if determinism is the case, so that every event is caused by some other event, then the chain of causes extends outside our selves and the individual agent is not the cause of his or her actions. On the other hand, if a person is a substance which has experiences, then this substance which lies outside the causal network can originate action without being caused to act by any event, or any other substance. When this happens we are free and we are responsible for our actions. Thus, the libertarian view avoids determinism, since some events are not caused by prior events, but it also avoids the strong form of indeterminism according to which some events occur uncaused, or occur for no reason whatsoever. Both freedom and responsibility are presented.

SOPHIA: That is a good summary of my position.

IVAN: Perhaps, but then I do not see how your view really avoids the problem facing the indeterminist; it just pushes the problem one step back. Admittedly, an action performed by a substantialist agent is not purely spontaneous, since it is caused by the agent who chooses it. But what of the agent's choice to bring about one action rather than another? Presumably that is an uncaused choice. How then can we hold an agent responsible for the actions he or she causes if the agent bringing about or choosing or causing one action rather than another is itself uncaused? If the doing of A and the doing otherwise are not causally determined, then either might occur given all the same past circumstances and laws of nature. But then we are left with the question: "Why did the agent do A rather than do otherwise?" If the agent's choice of A rather than B is just an arbitrary, spontaneous choice, then it would appear that it is not one for which the agent can be held ultimately responsible and the entire motivation for the libertarian view seems to collapse. How would you attempt to resolve this difficulty?

SOPHIA: Good question. I can just hint at the outlines of a solution now. I think what is necessary is some account of how there can be reasons for our final choice, without those reasons *determining* our final choice. Suppose a man is faced with an important choice. His girlfriend is pregnant and wants to give up the baby for adoption unless he agrees to marry her. He has to make a decision. Where does free will come in? Suppose that in this case there is real conflict. Given the man's character there is a pull toward both sides. He has reasons for marrying her, e.g., he wants to have a child, but he also has reasons for not marrying her, e.g., he is not sure that she is the person for him. To make a decision, an effort of will is going to be necessary. That is, whatever he chooses will involve an inner struggle and effort of will which must be exercised against opposing inclinations. The result of this effort is a decision to marry her or not marry her. The opposing desires and beliefs make an effort of will necessary to make a choice between one set of reasons rather than another. But the effort of will which terminates in the choice is not causally determined. In other words, his wants, desires and beliefs can provide reasons for either choice he makes, and yet these reasons do not determine his choice; he is capable of choosing differently. Thus his decision is free, and one for which he can be held responsible.

IVAN: I agree that sometimes we can give a teleological explanation of why people act by appealing to the end or goal they hope to achieve, but as I see it, if we can supply reasons for why people act the way they do, those reasons are still causal.

SOPHIA: And, of course, I disagree.

IVAN: Even so, our discussion certainly has helped to bring some of the central issues into clearer focus. Before going to class I want to touch on a point I suggested earlier, namely, that the ability to do otherwise should not be thought of as necessary for freedom and responsibility.

PHIL: What is your reasoning for that claim?

IVAN: It is based on a rather hypothetical situation, but sometimes a consideration of such situations is a useful technique to bring out the nature of a certain concept. Suppose, then, that a mad neuro-surgeon has created a machine that is attached to my brain. This machine, when turned on, ensures that whenever I want a piece of fruit I will choose an apple. Suppose further that I go to the farmer's market and having looked over the fruit I decide *of my own free will* that I want an apple. Following my want, I reach for an apple, buy it and begin to eat it. In this case, I couldn't have done other than buy an apple since the machine would have interceded and prevented me from buying a pear. But I didn't want a pear, I wanted an apple, and that is why we would not hesitate to say that my eating the apple was a free action even though I could not do anything else. We say this because I did not get any help from the mechanism, i.e., the mechanism played no role in my choice to buy an apple. Rather I decided to have an apple, and that is why the action is free; my being unable to do otherwise has nothing to do with it.

SOPHIA: The question of whether or not the "could not do otherwise" principle deserves our future philosophical allegiance is highly debatable. But since I have to teach, perhaps we can pursue that question further some other time.

IVAN: Sounds good.

NOTES

1 Clarence Darrow (1957) "The Crime of Compulsion," in Arthur Weinberg (ed.) *Attorney for the Damned*, New York: Simon and Schuster, reprinted in Victor Grassian (1984) *Perennial Issues in Philosophy*, Englewood Cliffs, NJ: Prentice Hall, p. 427.

GLOSSARY OF TERMS

Compatibilism There is human freedom and determinism is the case. This view is also called "soft determinism."

Hard determinism Since determinism is true we are neither free nor responsible for our actions.

Incompatibilism The view that human freedom and causal determinism cannot both be true. *See* libertarianism and hard determinism.

Initial conditions A set of circumstances which, together with laws, imply that a certain event will occur.

Law of nature A law is a generality (of the form, for every X if X has the property F_1, then X has the property F_2) that describes a relation between the properties of things.

Libertarianism Determinism is false, and therefore we are free and responsible for our actions.

Quantum mechanics A theory of the structure and behavior of particles. In this theory there is an element of unpredictability and randomness in the behavior of micro-particles because particles no longer have separate, well-defined positions and velocities.

STUDY QUESTIONS

1 What intuitions give rise to the problem of human freedom?
2 Evaluate Darrow's argument for hard determinism.
3 How could one argue that freedom and determinism are compatible?
4 What arguments does Sophia raise against the compatibilist position? How does Ivan attempt to answer them? Do you think any of his responses are adequate?
5 What connection is there between the libertarian conception of freedom and the substance view of personal identity?
6 Is it a necessary condition of one's having done an action freely that it could have been otherwise?
7 Do you think the compatibilist or the libertarian view of freedom is more plausible? Explain.

FURTHER READING

Anderson, Susan Leigh (1981) "The Libertarian Conception of Freedom," *International Philosophical Quarterly* 21: 391–404.

Discusses the connections between personal identity and freedom, and defends the libertarian view against criticisms.

Chisholm, Roderick (1964) "Human Freedom and the Self," Lindley Lecture, University of Kansas, pp. 3–15.

Chisholm argues that a solution to the problem of freedom involves incompatibilism and a substantialist conception of the self.

Double, Richard (1991) *On the Non-Reality of Free Will*, New York: Oxford University Press.

Defends the thesis that there can be no such thing as free will and moral responsibility.

Frankfurt, Harry G. (1971): "Freedom of the Will and the Concept of a Person," *Journal of Philosophy* 68: 54–68.

A classic article that argues against the ability to do otherwise as a condition of freedom.
Honderich, Ted (1993) *How Free Are You? The Determinism Problem*, New York: Oxford University Press.
*Kane, Richard (1989) "Two Forms of Incompatibilism," *Philosophy and Phenomenology Research* 50: 219–54.
A sophisticated defense of incompatibilism.

Appendix

Physical time and the universe

Quentin Smith

In Parts I to III of this book we have treated the topics of the finite and the infinite (Part I), time and identity (Part II) and the nature of freedom (Part III). These discussions have touched upon some issues that intersect with current scientific theories of time and change and it is useful to examine the relevance of the scientific theories to the debates. More importantly, there are a number of significant features of time and change that were not addressed in the dialogues and that deserve a separate discussion in the context of contemporary physical theories.

The most relevant scientific theories are Einstein's Special and General Theories of Relativity. Einstein's theories bear upon some of the issues discussed in the dialogues: for example, some philosophers argue that Einstein's Special Theory of Relativity shows that the tenseless theory of time is true and this argument needs to be evaluated. Further, certain types of change and global structures of the universe (a cyclically repeating history) require that the substantival theory of time be true and it is worth spelling out this argument for the substantival theory of time. Moreover, if time is "closed" (and has the structure of a circle), then time will neither begin nor end, but it will also be true that neither the past nor the future is infinite. These and other possibilities will be discussed in this Appendix. In particular, we shall concentrate on possible features of time and change that were not mentioned in the preceding dialogues. In the dialogues, we more or less assumed that there is only one time-series, that time either begins or has an infinite past, that time does not speed up or slow down, that the relations of simultaneity between any two events (e.g., two lightning-flashes) are the same for everybody, and the like. These ideas are consistent with the common-sense conception of time or with traditional Newtonian physics, but Einstein's theories force us to face other and more unusual possibilities. We learn from Einstein that it is possible that one and the same event is future for me and past for you. It is also possible for time to slow down and speed up, depending on the motion of a body. And time can be closed like a circle (time can start at point A, proceed to B and C and then return to A again, such that A, B, and C are arranged in a circle). It is possible for multiple time-series to exist; there can be times separate from the time in which we live, so that nothing in the separate time-series is simultaneous with, earlier than or later than, anything in our time-series. This would be the case if there were a number of universes distinct from and spatially and temporally unconnected to our universe. And events can occur in reverse order after first occurring in their natural order; events can occur in the order 1, 2, 3, 4, 3, 2, 1. We can first live our lives forwards, and then live them in reverse, going from an old person to a middle-aged person to a young infant. Furthermore, if our universe is rotating, then time-travel is possible; I can take off in a rocket and land on the earth at an earlier time, say in 1787.

These and other theories of time and change that are based on Einstein's theories shall be discussed in this Appendix. The preceding three parts of this book admitted of dialogue form, since they present arguments pro and con some given issue, e.g., compatibilist versus incompatibilist theories of free will, the tensed versus the tenseless theory of time, the relational versus the substantival theories of time and the like. However, we now come to an area that requires exposition (rather than argument pro and con on some given issue) and a dialogue form is unsuitable.

The Appendix has two main parts. In the first part, Einstein's Special Theory of Relativity is discussed and in the second part Einstein's General Theory of Relativity and contemporary cosmological theories are discussed.

Section A

Physical time in Einstein's Special Theory of Relativity

INTRODUCTION

Albert Einstein revolutionized the physical theory of time mainly by arguing that temporal relations (simultaneity, being earlier than, being later than) are not absolute but relative. A temporal relation is absolute if it is a two-termed relation. "The sun's rising over the earth is simultaneous with the sun rising over the planet Pluto" expresses a relation of simultaneity that is *absolute* in the sense that it is a relation between two items only, the two sunrises. A simultaneity relation is relative if it is a three-termed relation and one of the terms is *a reference frame* in relation to which two events possess the relation of simultaneity. A sentence that expresses a relative simultaneity is complex: "The sun's rising over the earth is simultaneous with the sun rising over the planet Pluto *relative to the reference frame of the earth*, but the sun's rising over the earth is earlier than the sun rising over Pluto *relative to the reference frame of a rocket passing by Mars*."

Einstein argued not only that temporal relations are relative, but also that the *length of temporal intervals* is relative to a body's motion, as measured by some reference frame. The faster a body travels relative to some reference frame R, the slower the time passes on the body, as measured by the reference frame R.

In order to see how Einstein arrived at these novel results, we must begin with his rejection of absolute space and his incorporation of the speed of light into his physical laws.

THE THEORY OF ABSOLUTE SPACE

Isaac Newton supposed that there is absolute space. Absolute space is a rigid and unmoving container of all bodies. Absolute space is the single reference system in terms of which real motion and rest are defined, and in relation to which bodies possess their real lengths. According to Newtonian science, "The earth is moving at 65,000 miles an hour" means "The earth is moving at 65,000 miles an hour through absolute space." And "A stick is twelve inches long" implies "A stick would be measured to be twelve inches long if it were measured while it was at rest in absolute space."

However, as Newton was aware himself, the motion, rest and length of bodies we actually observe are motions, etc., relative to some other body. For example, if we observe a train to be moving, we observe it to be moving relative to the earth. We observe the earth moving relative to the sun, the sun moves relative to the center of our galaxy, the Milky Way galaxy, which in turn is observed to move relative to other galaxies. The question that naturally arises at this point is, *how do we tell if a body is moving or at rest in absolute space*?

Newton believed it is meaningful to speak of absolute motion and rest, not just motion and rest relative to this or that body, the earth or sun, even if we cannot observe absolute motion or rest. For this reason, his laws of physics referred to absolute motion and rest. For example, his first law of motion states that *In the absence of outside influence, a resting object will remain at rest, and a moving object will remain in motion in a straight line at constant speed.* By "at rest" Newton meant "at rest in absolute space" and by "in motion in a straight line" he meant "moving in absolute space in a straight line."

However, the idea that the laws of physics should refer to something *in principle unobservable* does not sit well with many physicists and they searched for a way to observe absolute motion and rest. In 1885, two scientists, Michelson and Morley, believed they had devised an experimental apparatus that could detect the earth's motion in absolute space. This became the most famous experiment in the history of science, for it caused the downfall of Newtonian physics and laid the foundation for Einstein's Special Theory of Relativity. It is called the Michelson–Morley experiment.

Michelson and Morley tried to measure the velocity of the earth through absolute space. The idea was to measure the velocity of the earth through *the ether*, which is supposed to be at rest in absolute space. Just as air is the medium through which sound-waves travel, so ether was thought to be the medium through which light-waves travel.

If the earth moves through the ether, we should expect the earth to move through a wind, an ether wind, much as a car traveling through the air travels through an air wind (as you can tell by putting your hand out the window). Michelson and Morley believed *the velocity of the earth through the ether wind* could be measured. Since the ether is at rest in absolute space, this would give us the velocity of the earth with respect to absolute space.

Michelson and Morley used two mirrors to measure the velocity. They sent light pulses to the mirrors and measured how long it would take for the light pulses to go from a mirror back to their source. The velocity of the light should be different in different directions, since a light-ray projected against the ether wind should be slightly retarded by the ether wind, just as a swimmer is retarded when swimming upstream. And a light-ray projected with the ether wind should have its velocity increased by the ether wind, just as a swimmer's speed increases as she swims with the current.

Now light travels at 186,284 miles a second. Since the earth is moving at 18 miles a second in its orbit around the sun, a light-ray traveling *against* the ether wind should go at 186,284 *minus* a few miles a second. A light-ray traveling *with* the ether wind should travel at 186,284 *plus* a few miles a second.

But Michelson and Morley measured light to be traveling at 186,284 miles a second, *regardless of the direction of the light-ray and regardless of the position of the earth*. The idea that the earth is permanently at rest in absolute space is ruled out as a possible explanation, since the earth is in elliptical motion around the sun.

But how, then, should we explain the result that light is always measured to be traveling at 186,284 miles per second? Einstein's answer was that this fact does not need to be explained in terms of anything deeper, but is a fundamental postulate of physics.

THE TWO POSTULATES OF EINSTEIN'S SPECIAL THEORY OF RELATIVITY

The reasoning underlying Einstein's Special Theory of Relativity has the form of *modus tollens*:

If P, then Q.
Not-Q.

Therefore,

Not-P.

Applied to absolute space and light, the argument goes:

1 If there is absolute space, then there is a privileged reference frame, at rest in absolute space, in relation to which light is measured to have its true velocity of 186,284 miles per second, and bodies moving in absolute space measure light to have a higher or lower velocity.
2 It is not the case that some bodies measure light to have a velocity different than 186,284 miles per second; every body, regardless of its state of rest or motion, measures light to have this value.

Therefore,

3 It is not the case that there is absolute space.

Einstein rejected the idea of absolute space and the idea of there being a privileged reference frame; every reference frame is equally valid and light is measured to have the same velocity in each reference frame. This led to the two fundamental axioms of the Special Theory of Relativity. These two axioms are *the Principle of Relativity* and *the Principle of the Constancy of the Velocity of Light*. As they are stated in Einstein's original 1905 paper, "On the Electrodynamics of Moving Bodies," they read:

> the same laws of electrodynamics and optics will be valid for all frames of reference for which the equations of mechanics hold good. We will raise this conjecture (the purport of which will hereafter be called "the Principle of Relativity") to the status of a postulate, and also introduce another postulate, which is only apparently irreconcilable with the former, namely, that light is always propagated in empty space with a definite velocity c which is independent of the state of motion of the emitting body.[1]

The significance of these two postulates and their implications for the theory of time are not apparent at first glance and require further unpacking. Let us consider first the implications of the postulate about light.

THE STRANGE BEHAVIOR OF LIGHT

The following three principles cannot all be true:

1 The laws of physics are the same for all frames of reference.
2 The velocity with which light travels is constant, regardless of whether the body is moving towards or away from the source of the light.
3 Newton's principle of the addition of velocities.

Einstein based his Special Theory of Relativity on the idea that (1) and (2) are true and (3) is false. This requires that we recognize that the behavior of light is very strange indeed.

The reason why light's behavior is strange is that Newton's principle of the addition of velocities is the straightforward principle that we all apply in our daily life. According to this principle, if a person is walking forwards in a train at 2 miles an hour and the train is traveling at 40 miles per hour with respect to the earth, the person's velocity with respect to the earth is 2 + 40 miles per hour.

This gives us one of Newton's transformation equations, which implies that to obtain the velocity of the person relative to the earth, you add up the velocity of the person relative to the train and the velocity of the train relative to the earth:

$$V^1 = U + V$$

In terms of our example, V^1 is the velocity of the person relative to the earth (42 miles an hour). U is the velocity of the train relative to the earth (40 miles an hour). V is the velocity of the person relative to the train (2 miles an hour). This equation seems at first glance to be obvious; its solution for our example is 42 = 40 + 2.

Newton's principle of the addition of velocities, however, is inconsistent with the results of the Michelson–Morley experiment. To understand this in more imaginatively vivid terms, consider the following facts about light.

Suppose you are traveling towards the sun at the velocity V and a ray of light emitted from the sun is traveling towards you at the velocity C (186,284 miles per second). If Newton were correct, you should measure the velocity of the light to be V + C. But as Michelson and Morley have shown, you do not; the velocity is still measured to be C.

Although physicists have become used to this behavior of light, it will strike those who are less familiar with it as resembling something that we might expect to meet in a nightmare. Suppose that you are at rest on the earth and measure light from the sun to be traveling past you at 186,284 mps. You then begin moving toward the sun at 185,000 mps. What happens? You would still measure light to be traveling past you at 186,284 mps. Suppose further somebody is traveling away from the sun, in the opposite direction, at 185,000 mps. That person will measure the same light to be traveling past her at 186,284 miles per second.

It also implies this. Suppose you are standing still and a beam of light passes you by at 186,284 mps and then you try and catch up to it. You have a spaceship that travels 186,274 mps. So you speed after the light pulse at 186,274 mps. Does this mean the light pulse is now receding from you at only 10 miles per second? No, it is still receding from you at 186,284 mps. You can never become closer to it, no matter how fast you go (the speed of light can never be reached by a massive body). This is indeed like a nightmare, or at least like a dream, where the ordinary principles of everyday life no longer apply.

One of Einstein's main ideas was to include this constant velocity of light in his mathematical equations about velocities; this inclusion enabled the correct observational results to be deduced. The fact that if you were traveling towards the sun at velocity V, you would still measure light to have the same old velocity, C, follows from one of Einstein's equations. Einstein replaced Newton's principle of the addition of velocities with his own principle:

$$V^1 = \frac{U + V}{1 + \frac{UV}{C^2}}$$

C is the symbol used for the velocity of light, 186,284 miles a second. This new equation embodies the experimentally confirmed postulate that the velocity of light is always measured to be the same, regardless of your state of motion relative to the source of the light. Let V^1 be the velocity of light relative to you as you travel toward the sun:

$$V^1 = \frac{V + C}{1 + \frac{VC}{C^2}} = \frac{V + C}{\frac{C + V}{C}} = C$$

In short, $V^1 = C$, that is, the velocity of light relative to you as you travel at any velocity V toward the sun will always be C.

This equation explains the results of the Michelson–Morley experiment. But what does this have to do with the nature of time?

THE RELATIVITY OF THE LENGTH OF TEMPORAL INTERVALS

By rejecting the principle of the addition of velocities, Einstein rejected two seemingly self-evident assumptions about space and time upon which they are based. Newton's principle of the addition of velocities is based on this false assumption:

1 Spatial lengths and temporal durations have a fixed physical value, regardless of the relative motion of the body with the measured length or the relative motion of the body with the measured duration.

However, Einstein's principles implied that spatial and temporal lengths and relationships are different for relatively moving systems. I will explain this implication for temporal intervals.

An example shows why time goes slower on a body that is moving, relative to some other body. Suppose a rocketship is traveling past the earth at a great velocity. A person on the earth has a light clock. It takes one unit of time for the light to go from the bottom to the top of the clock, and a second unit of time for the light to go from the top to the bottom.

The person on the rocket has an identical light clock and this light clock can be observed by the person on the ground and compared with his own light clock. The person on the ground observes the rocket's light clock and observes it to be "ticking" more slowly that his own. This is because the rocket is moving and the light pulse has a longer distance to travel from the point of origin to the top of the light clock (see Figure 1). Note that the fact that the rocket is moving does not mean that the person on the earth would measure the light to be traveling at a *different velocity* inside the rocket's light clock. Einstein's postulate of the constancy of the velocity of light implies that the earth-bound

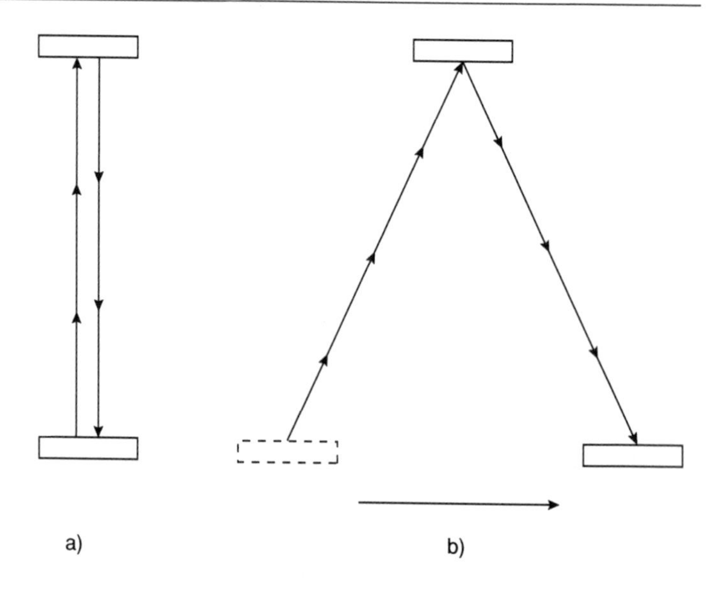

Figure 1 Light clock
a) Earth Clock as seen from earth.
b) Rocket Clock as seen from earth. The rocket is moving in the direction of the horizontal arrow. The light signal is sent off at the same time as the light signal on the earth clock is sent.

person would measure the rocket's light to have the same velocity C as the light in his own light clock.

Whereas the single event of the light going from the bottom to the top of the clock and back to the bottom in the earth-bound person's light clock takes two time-units, this same person judges it to take three time-units in the rocket's light clock. Time goes slower in the rocket than it does on the earth, as measured by the person on the earth.

Exactly the reverse is the case, relative to the reference frame of the rocket. The person in the rocket may take herself to be at rest, and to regard the earth as in motion. The person in the rocket will judge the time on the earth to be slower than her rocketship time. This is the basic idea behind the thesis that the *length of temporal intervals is relative to a reference frame.*

Since there is no absolute space, no body is absolutely at rest (i.e., at rest in absolute space) or absolutely in motion. Motion and rest are relative. Thus, the question, "Who is really at rest, the person on the earth or the person in the rocket?" is a meaningless question. Both can correctly regard themselves as at rest and the other person as moving relative to themselves.

A true example of a difference in temporal intervals may be given. If a spaceship travels by the earth with a light clock and the spaceship is traveling at 99.5 per cent of the speed of light, nearly 186,000 mps, then a second hand on a clock in the spaceship would move only six seconds during the time that the second hand on our clock would move sixty seconds. Time goes ten times slower on the spaceship, relative to the reference frame of the earth. Every physical process on the spaceship goes slower, relative to the reference frame of the earth. Thus, people on the rocketship would age slower and live longer than the people on the earth, relative to the reference frame of the earth. Conversely, the people on the rocketship would observe people on the earth aging more slowly and living longer.

THE RELATIVITY OF SIMULTANEITY

Einstein's two postulates of the constancy of the velocity of light and the relativity of the laws of nature to a reference frame also imply that the temporal relations of simultaneity, earlier than and later than, are relative. As I explained at the beginning of this essay, "relative" and "absolute" have a special meaning as applied to these temporal relations. A temporal relation is absolute if it is a two-termed relation. "A is simultaneous with B" expresses a relation of simultaneity that is absolute in the sense that it is a relation between two items only, A and B. A temporal relation is relative if it is a three-termed relation and one of the terms is *a reference frame* in relation to which two events possess the relation of simultaneity, earlier than or later than. "A is earlier than B relative to the reference frame R" expresses a temporal relation that is relative.

The simplest way to explain the relativity of simultaneity is to use an example Einstein gave, involving a hypothetical train that is traveling at a velocity of over half the speed of light. Suppose there is a train traveling in the direction of A (see Figure 2).

Two lightning-flashes hit the front and back of the train and scorch the ground. We will suppose there is a person Nadia

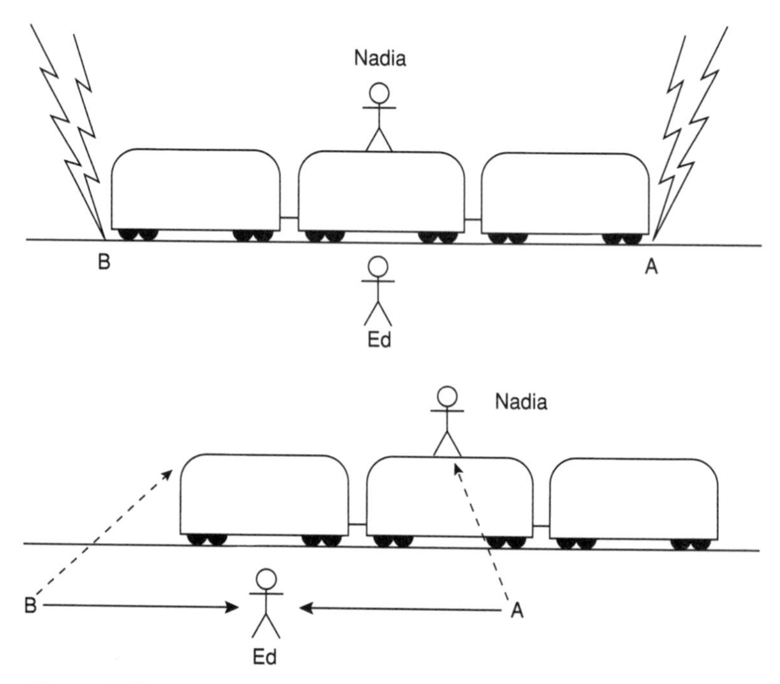

Figure 2 Einstein's Train

sitting on the top of the train in the middle and another person Ed sitting on the ground, half-way between points A and B.

Ed on the ground receives the lightning-flashes simultaneously, so he judges the two strikes to have hit the points A and B simultaneously. But Nadia disagrees. She is on the train and the train is moving toward the point A where one flash struck, and is moving away from point B. So the flash-strike A reaches her first and she judges the flash to have hit the front of the train before it hit the back of the train.

We would normally assume that Nadia is in motion and that Ed is at rest. We would say that she is on a train that is moving toward the point A and that this explains why the reflection from the flash-strike at A reaches her first. We say that once she makes allowance for the train's movement, she will realize that A and B occurred simultaneously, just as Ed on the ground realized this.

However, Einstein's theory implies that it is false that Nadia is in motion and Ed at rest. They both can equally well regard themselves as at rest and the other person in motion. Nadia can regard the train at rest and the earth as moving past the train. This explains why, from her vantage point, Ed receives the flashes simultaneously; he is moving toward the later flash B and away from the earlier lightning-strike A – and so B reaches him at the same time as A.

It is important to explain that the temporal relations implied by Einstein's Special Theory of Relativity are not in every case relative relations. Einstein's theory implies that some events are related by absolute temporal relations.

ABSOLUTE TEMPORAL RELATIONS IN THE SPECIAL THEORY OF RELATIVITY

Some events sustain absolute (two-termed) relations of simultaneity, earlier than or later than. Consider my birth and my speaking now. My birth is in some indirect sense a cause or at least causal precondition of my speaking now as an adult. Clearly, it would not make sense for my birth to be simultaneous with my adult talking phase, or later than this phase, relative to some reference frame. Rather, my birth is earlier than my adult talking phase relative to every reference frame. More simply, my birth is earlier than my talking now in an absolute sense. Generally speaking, if something C is *a cause* of something else E, then C is *absolutely earlier* than E.

The absolute temporal relations postulated by the Special Theory of Relativity are best explained in terms of the notions of a *light cone* and and a *space-time diagram*. A space-time diagram of a light cone appears in Figure 3.

In Figure 3, the direction of time, from earlier to later, is from bottom to top (see the time arrow). Thus, event A is later than event C, and events J, D and F are simultaneous. Distances in space are represented by the horizontal line. So A, D and C occur at the same place (but at different times) and events J, D and F occur at different places (but at the same time).

Regarding the causal relations, event C causes event D and event D causes event A. This implies that C is absolutely earlier than D and D absolutely earlier than A. Events that are causally connected are related by absolute temporal relations. Events that are not causally connected are related by relative temporal relations.

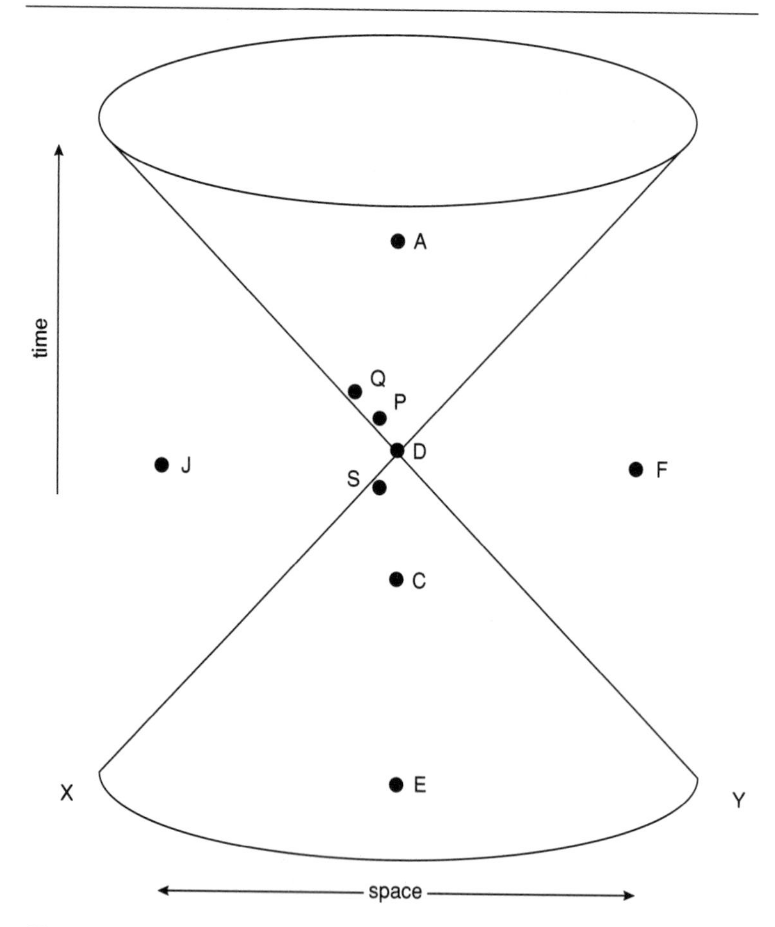

Figure 3 Light cone

The lines that constitute the "light cone" may be explained in terms of an example. Suppose the event D is the event of my speaking now. The lines X to D and Y to D represent light signals or flashes sent from past events X and Y to D, my speaking. Let us say they are two stellar explosions that occurred billions of miles from here at different places and that traveled toward my eyes, so that I can see them both now, at event D.

The event C and all events in the bottom cone between X and Y are events that can be connected to my speaking now at D by signals, transmissions or other sorts of causal processes that travel below the speed of light. For example, air molecules in a blowing wind that hit my mouth travel slower than the speed of light and would causally influence my speaking at D.

Now consider the top half of the light cone. These contain events in the future of my speaking, in my absolute future. Whatever event my speaking can causally influence is in my absolute future. The sound-waves leaving my mouth and entering your ears (at event P) occur in my absolute future, so the event P consisting of your hearing what I am saying is in my absolute future. Also in my absolute future is somebody further away from me who hears indistinct and faint noises coming from my mouth; her hearing these noises (at event Q) is further in my future than your hearing since it takes longer for the sound-waves to travel to her.

The lines that represent the boundaries of the light cone correspond to light-rays that arrive at event D or that leave event D. Nothing can go faster than light and so the paths of light naturally represent the boundaries of what can causally influence me from the past and what I can causally influence in the future.

Consider further the event of my birth, which we may represent as event E in our diagram, which is even further in the absolute past of D than is event C. The event E of my birth has a remote causal influence on my speaking. My birth causally affected my immediately subsequent bodily state, that affected my next bodily state, and so on up until events C and D. For example, the beating of my heart now at event D is a remote effect of my birth; there is a causal chain of events in my body leading from my birth to my present bodily state.

So far I have discussed the bottom half of the light cone, which is the absolute past of event D (consisting of everything that can causally affect D) and the top half of the light cone, which is the absolute future of D (consisting of everything that D can causally affect). It remains to discuss the area in the diagram outside the light cone.

This area represents all the events that cannot causally influence the event D of my speaking and that my speaking cannot causally influence. For example, if a meteorite impacts on the planet Uranus as I speak, at event J in the diagram, this impact cannot causally influence my speaking. Nothing can go faster than light and it takes several hours for light from the impact on Uranus to reach the earth, and by that time I will have finished speaking. Likewise, I cannot send a light signal now and have it reach Uranus during the time of the impact.

The event J of the meteorite impact is *causally unconnectable* to the event D of my speaking. All events outside the light cone are causally unconnectable to D. The significant fact about these events is that *only these events* sustain *relative* temporal relations to D. More exactly, these events are connected to the event D by the absolute relation of *being causally unconnectable* to D, but in addition they are connected to D by a variety of relativistic temporal relations involving simultaneity, being earlier than and being later than.

An example makes this clear. Suppose there are two observers who pass through event D, one being O_1 and the other O_2. O_2 is moving relative to O_1, so her path in space–time is an angle, as in Figure 4. Here the lines represent the "world-lines" of the two observers O_1 and O_2. A world-line is the path of a thing through space and time. It is the successive positions in space that a thing occupies. First O_1 is at C, then at D, then at K and then at A.

The broken lines are the planes of simultaneity of O_1 and O_2 when they are located at the event D. Every event on the plane of simultaneity S_1 is simultaneous with the event D, as judged by O_1. Every event on the plane of simultaneity S_2 is simultaneous with the event D, as judged by O_2.

A rather dramatic story may be used to illustrate the relativity of the time-relations between D and the events that fall outside of D's light cone, such as the events F and G. (D's light cone is not drawn, but as can be inferred from Figure 4, the events F and G are outside of D's light cone.) Suppose O_1 and O_2 are sisters and have the same biological parents and that event G is the event of their parents dying in a car crash. As the two sisters pass each other at event D, one is an orphan and the other's parents are still alive. This is possible since event G is outside event D's light cone. The sister O_1 determines event F to be simultaneous with D, and G to be later than D, but O_2 determines G to be simultaneous with D, and F to be earlier than D.

The question may be raised as to how O_1 and O_2 can know that F or G is simultaneous with the event D, considering that F and G fall outside of D's light cone. There are no causal connections between F and D and between G and D. So there must be some special method introduced by means of which O_1 and O_2 can know that F or G are simultaneous with D.

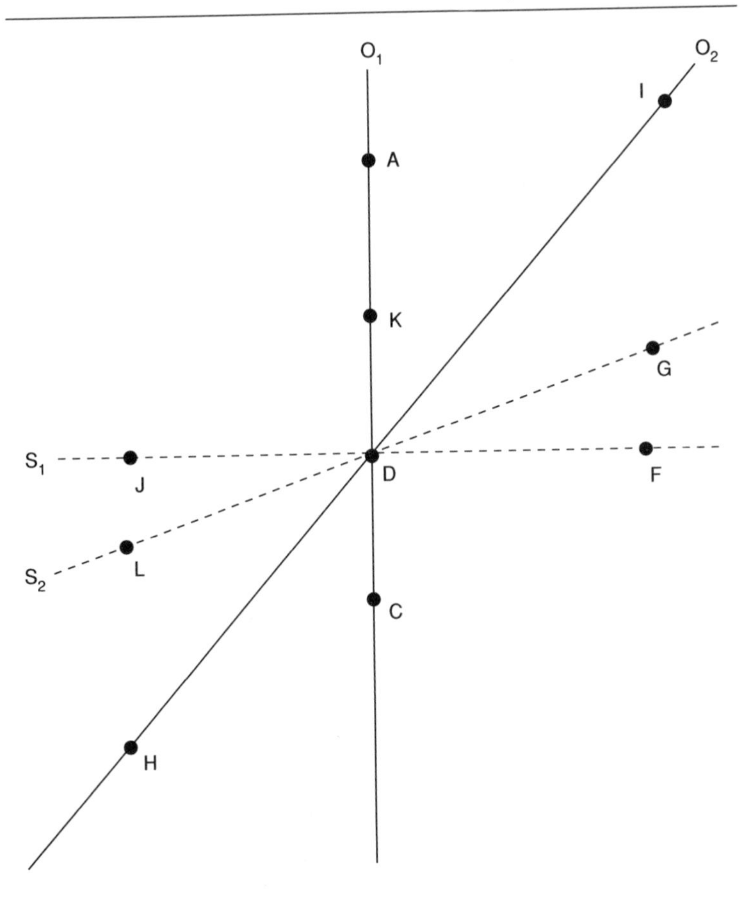

Figure 4 Two world-lines.
O_1 is O_1's world-line
O_2 is O_2's world-line
S_1 is O_1's plane of simultaneity at D
S_2 is O_2's plane of simultaneity at D

The method in question is the Light Signal Method. This involves bouncing a light signal off a distant object and measuring the time which elapses between transmitting the signal and receiving the reflection. Suppose I send out a signal at 2 o'clock and receive it back at 4. The light signal comes back and gives me information of a car crash; the light signal bounced off the car crash. Since the light signal took one hour to go out and one hour to return, I conclude that the car crash from which it reflected occurred one hour later than 2 o'clock and one hour earlier than 4 o'clock; it occurred at 3 o'clock.

Thus, O_1 and O_2 learn *at a time later than D* about what is simultaneous with D. Since F and G are outside of D's light cone, O_1 and O_2 cannot know at D what is simultaneous with D; but at a later time, say at the event K for the observer O_1, she is in a position to receive light signals she earlier had transmitted to F and can determine then which events were simultaneous with D. Suppose D is half-way between events C and K. D occurs at 3 o'clock. The observer O_1 sends out light signals towards the events F and J at an earlier time, at the event C, when it is 2 o'clock. One of the signals bounces off the event F, which is the event of her parents climbing into their car. She receives the signals back at event K, when it is 4 o'clock. So she concludes that at 3 o'clock her parents were still alive. (And suppose event J is the event of a leaf falling elsewhere; she concludes that her parents getting into the car, D and the leaf falling were simultaneous.)

At the event C, O_1 also sends out signals towards the event G, which (as she learns at event A, at 5 o'clock) is the event of the car crashing and her parents dying. Since G is near in time to K (as is indicated by Figure 4), she concludes that G occurred close to 4 o'clock and that her parents were alive at 3 o'clock.

By contrast, her sister O_2 draws a different conclusion about the time of the car crash. On O_2's world-line, the event D is half-way between events H and I. O_2 sends out light signals at event H in the direction of events L and G and she receives their return reflections simultaneously at event I. She concludes that L, D and G are simultaneous. For O_2, the car crash and her parents' death occur at the same time D occurs.

There is an analogy between this situation and the train example. The event D at which O_1 and O_2 are at the same place at the same time is similar to the event in the train example when Nadia on the train top is exactly opposite to Ed on the bank, as the train passes (see Figure 2). Of course, "at the same place at the same time" is meant approximately, since Ed and Nadia cannot occupy the identical place at the same time. Although Ed and Nadia are at the same place at the same time, they come to different conclusions about whether the lightning strike at the front end of the train is simultaneous with the lightning strike at the rear end of the train. This is due to the fact that they are in motion relative to each other and there is no absolute state of rest or motion by which they can determine who is "really" right about the time of the two strikes. Analogously, since neither O_1 nor O_2 can determine who is "really" at rest (motion and rest being merely relative), neither O_1 nor O_2 can be said to have the "right answer" about whether D is simultaneous with the car crash at G or whether D is earlier than G.

THE TENSED THEORY OF TIME AND EINSTEIN'S SPECIAL THEORY OF RELATIVITY

Einstein's Special Theory of Relativity has seemed to some to have implications for the debate between the defenders of the tensed theory of time and the defenders of the tenseless theory of time. Some philosophers and physicists argue that Einstein's theory implies the tenseless theory of time. They argue it implies that future events are equally as real as present events, and therefore that the tensed theory that future events are not-yet-real is false. For example, it might be argued that since the car crash is in O_1's future *and in* O_2*'s* present, then it must exist, it must be fully real.

A better statement of this detenser argument can be given in terms of a different example. We may modify Figure 4 by omitting O_1's world-line and inserting a new world-line O_3, which is the world-line of the father of O_1 and O_2 and who dies in a car crash at G (see Figure 5). The father's world-line extends through the event F (his getting into the car) and ends at G with his death. Suppose F is present for O_3. The father is just getting into his car. The event G of his death in the car crash is in his future; it is not yet real. Since D (the event of his two daughters passing each other) is simultaneous with F for O_3, D is also present for O_3.

Suppose further that D is present for his daughter O_2. But the interesting fact is that the event G of the car crash is on O_2's plane of simultaneity at D; when D is present for O_2, her father's death is present. If it is present for her, it is real for her. O_2 will at a later time (at the event I) receive light signals from the car crash, which will prove that the crash was real when she was located at the event D. This is based on the Light Signal Method: O_2 sends signals out at H and receives them back at I; since D is half-way between I and H, D is simultaneous with whatever reflects back the signals (in this case the car crash G).

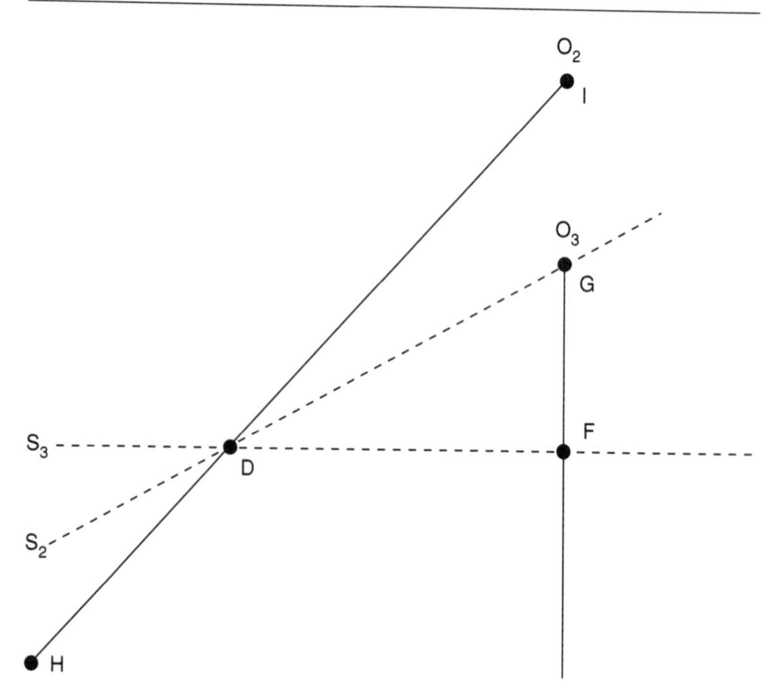

Figure 5 World-lines of O_2 and O_3

The fact that G is real for O_2 when it is future for O_3 can be demonstrated more dramatically if we change the example somewhat and suppose that the father O_3 survives the car crash and his world-line continues on and intersects O_2's world-line at the event I. Suppose O_2 took a flash photograph of the crash at G. She pressed the picture-taking button on the camera at the event H, the light signals from the camera bounce off the car crash when she is at D, and the signals return to her camera and produce a photograph at event I. So O_2 can show her father at I a photograph that proves to him that his car crash was real when his event F of getting into the car was present.

The next step in the argument for the tenseless theory of time is to argue that G is fully real for the father O_3 when he is at the earlier event F. This is argued as follows. When O_3 is getting into his car at F, F is fully real for him. Since the event D is simultaneous with the event F, D is also fully real for O_3. Now G is also on a plane of simultaneity with D, so G is fully real for O_3 if D is fully real for him. This is based on the principle that if an observer at D is fully real for O_3 at F, then whatever is fully real for the observer at D is also fully real for O_3 at F. It follows that when F (the event of his getting into the car) is fully real, G (his death in the car crash) is fully real, even though it is later on O_3's world-line.

The proponent of this argument believes it shows that the tenseless theory of time is true. According to the tenseless theory, all events are fully real; they are all equally real, and so if one event is fully real, so are all other events. The tensed theory, by contrast, holds that the future is not-yet-real and so is unlike the present event, which is fully real. Since the Special Theory of Relativity is a well-confirmed scientific theory and implies that future events are fully real, it follows (according to the proponents of this argument) that the tensed theory of time is false.

What shall we make of this argument? The most important thing to note is that the argument depends on the principle that "If an observer at D is fully real for O_3 at F, then whatever is fully real for the observer at D is also fully real for O_3 at F." This principle does not follow from Einstein's Special Theory of Relativity but is a metaphysical assumption adopted by the proponents of the argument. The defender of the tensed theory of time need not accept this principle. Indeed, the defender of the tensed theory of time will hold that reality is relative in the very same respect that presentness is relative. For O_3 at F, F has presentness and D has presentness, but the car crash G has futurity. There is no need to assume that if G is fully real for O_2 at D, then it is also fully real for O_3 at F. We do not assume this about presentness; we do not assume that if O_2 at D has presentness relative to O_3 at F, then anything that has presentness for O_2 at D *also* has presentness for O_3 at F. We assume that the event G, the car crash, is future for O_3 at F. Since reality depends on what is future, present and past, we should assume that if something is future for O_3 at F, then it is not-yet-real for O_3 at F.

The defender of the tensed theory of time will point out that Einstein's Special Theory of Relativity may be taken as implying that *reality* is relative to an observer. This will not be especially troublesome to the tenser, since he or she holds that *being fully real* and *being present* are logically equivalent; if presentness is relative, and reality consists of what is present, then it follows that reality is relative. The tenser will argue that this thesis is no more implausible than the theses that time, distance, length, etc., are relative to reference frames, which Einstein's theory implies irrespective of the differences between the tensed and tenseless theories of time.

The upshot of this discussion is that Einstein's Special Theory of Relativity does not show that the tensed theory of time is false. Einstein's theory is consistent with both the tensed and the tenseless theories of time and which of these two theories is the correct one needs to be decided by philosophical argument. In a word, this is a metaphysical issue, not a physical issue.

NOTES

1 Albert Einstein *et al.* (1952) *The Principle of Relativity*, New York, pp. 37–8.

Section B

Physical time in current cosmologies

THE GENERAL THEORY OF RELATIVITY

Einstein's aim in constructing his General Theory of Relativity was to develop a theory of gravity that is consistent with the results of his Special Theory of Relativity. This required Einstein to reject Newton's theory of gravity. Newton's theory implied that the force of gravity is propagated at an instantaneous or "infinitely fast" speed, which is inconsistent with Einstein's postulate that the speed of light (186, 284 miles per second) cannot be exceeded. Further, Newton's theory implied that the events connected by the instantaneous force of gravity were absolutely simultaneous, but Einstein's Special Theory of Relativity implied that simultaneity is relative.

Einstein's point of departure was to reject the idea that gravity is a force, instantaneous or otherwise, and to postulate that gravity is the *curvature of space–time*. The extent and manner to which space–time is curved is dependent on the amount and distribution of the mass-energy in that space–time. The basic equation of the General Theory of Relativity, in a simplified form, reads

> (curvature of space–time) = 8(density and pressure of mass-energy)

This equation says that the degree and kind of curvature possessed by space–time is dependent on the density and pressure of the mass and energy in that space–time. If there is no mass-energy in space–time, then space–time will be flat (the curvature will be zero). But if there is some, then space–time will be curved to some degree, either on a global scale or locally around the concentrations of mass or energy. If the density and pressure of the mass-energy in the universe is sufficiently great, space–time as a whole will curve into a sphere. Different amounts and distributions of the mass-energy will result in different configurations of space–time. It is an empirical question as to which sort of configuration our universe possesses. It appears that our universe has a configuration that admits of a *cosmic time*.

COSMIC TIME

"Cosmic time" is the phrase used by physicists to denote the time of the universe or cosmos as a whole as it is represented in the General Theory of Relativity. Although philosophers usually confine their discussion of physical time to time as it is represented in the Special Theory of Relativity, this concentration is misplaced. The Special Theory of Relativity is a theory of physical phenomena that does not take into account the effects of gravity. This means that the Special Theory corresponds only to a universe in which the effects of gravity are non-existent, which would be the case only if there were no matter or energy in the universe. The Special Theory of Relativity describes a completely empty universe. Since our universe is not empty, the Special Theory does not describe our universe. It follows that the theory of physical time embodied in the Special Theory does not describe the physical time of our universe. The importance of the Special Theory lies in the fact that it contains ideas that are incorporated into the General Theory of Relativity, which does describe the physical time of our universe.

The theory of cosmic time borrows from the Special Theory of Relativity the notion of a plane of simultaneity. As I explained in the previous essay, a plane of simultaneity is determined by the Light Signal Method. If I send out light pulses at 2:00 p.m. and receive them back at 4:00, then every event from which the light pulses were reflected constitutes a set of events that are simultaneous with my location at 3:00 p.m. The events in this set constitute a *plane of simultaneity* that runs through the event consisting of my spatial location at 3:00 p.m.

We may define a plane of simultaneity more formally. Consider a world-line W_1 of a possible observer who exists through the successive times T_1, T_2 and T_3 (see Figure 6). A world-line is a succession of space–time positions that can be occupied by a particle. Any space–time position or event D (in relativity theory,

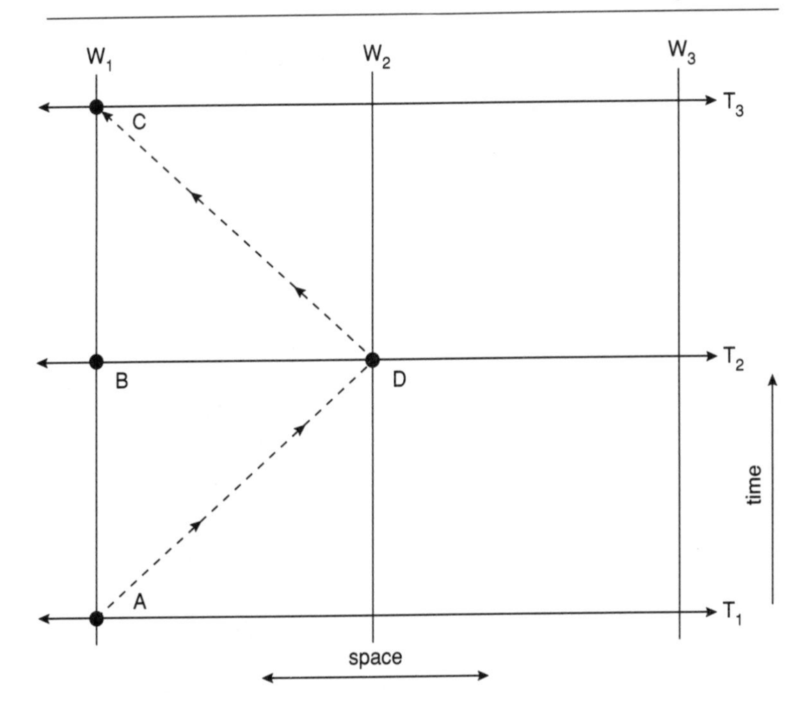

Figure 6 Cosmic time
W_1, W_2 and W_3 are world-lines of possible observers that are at rest relative to each other and are persisting through the successive times T_1, T_2 and T_3. T_1, T_2 and T_3 are planes of simultaneity that constitute part of a cosmic time-series. Light is sent from event A to event D and is reflected back to event C. D is simultaneous with the event B. Thus, D is on the plane of simultaneity T_2 that intersects the world-line W_1 at the event B.

space–time positions are called "events") is a part of the plane of simultaneity of the world-line W_1 at time T_2 if the following two conditions are met. (1) It is possible that at time T_1 a light signal is sent from A to the space–time position D, such that the light signal reflects off D and rearrives at the world-line W_1 at the position C. (2) The interval of time that elapses between the emission of the signal at A and the rearrival of the signal at C is twice as long as the temporal interval that extends from A to B and is also twice as long as the temporal interval that extends from B to C.

A *cosmic plane of simultaneity* is a set of interlocking planes of simultaneity of various world-lines W_1, W_2, W_3, etc., such that the interlocking planes of simultaneity are maximally extended, forming a universe-wide plane of simultaneity. T_1 and T_2 in Figure 6 are both cosmic planes of simultaneity. The sequence of all the cosmic planes of simultaneity that are parallel to T_1 and T_2 and that are either earlier or later than T_1 or T_2 constitutes a cosmic time-series. Since T_3 in Figure 6 is a plane of simultaneity that is parallel to T_2 and later than T_2, it belongs to this cosmic time-series.

There are many different ways of dividing space–time into different cosmic time-series; there is a different cosmic time-series for each set of parallel world-lines. There is some world-line W that is at an angle to W_1, W_2 and W_3, and W and all the other world-lines that are parallel to W have planes of simultaneity that constitute a different cosmic time-series. The fact that there are many different cosmic time-series, one time-series relative to each set of parallel world-lines, embodies *the relativity of temporal relations* that is a core feature of the Special Theory of Relativity.

One of the most interesting aspects of the notion of time in the General Theory of Relativity is that in some cases it allows that one of the many cosmic time-series constitutes a *privileged* time-series that deserves the title "*the* time-series of the universe." This is of particular interest since our universe meets the condition of having a privileged time-series and this series constitutes (in the final analysis) the actual referent of the phrase "the physical time of our universe."

The conditions of there being a privileged cosmic time-series are that the universe should be *homogeneous* and *isotropic*. If the matter in the universe is evenly distributed, the universe is said to be homogeneous. If the universe looks the same in every direction from each point in the universe, then the universe is said to be isotropic. On a very large scale, when we consider the super-clusters of galaxies (the clusters of clusters of galaxies) our universe is homogeneous and isotropic.

The universe-wide planes of simultaneity that coincide with the *planes of homogeneity* constitute the privileged cosmic time-series. A plane of homogeneity is a plane of space–time positions (maximally extended in both directions) where the density of matter is constant, the pressure of matter is constant and the curvature of space–time is constant. The universe would appear homogeneous and isotropic to any observer whose world-line possessed planes of simultaneity that coincided with the planes of homogeneity.

The reason that homogeneity and isotropy make a set of

parallel world-lines and their planes of simultaneity privileged is that it is *simpler* to formulate a scientific theory from the reference frame of these world-lines than from the reference frame of any other set of parallel world-lines.

In our universe, which is now expanding, the privileged world-lines coincide with *the uniform expansion of the universe*: the expansion of the universe appears to be uniform and regular from the reference frame of these world-lines.

The planes of simultaneity of the earth's world-line coincide approximately with the planes of homogeneity. This means that the time relative to the reference frame of the earth is approximately the same as the privileged cosmic time.

The laws of nature and the observational data that are formulated in astronomy and cosmology use the privileged cosmic time-series to date the relevant physical events and processes. For example, when it is said "The earth was formed 4.5 billion years ago," this means "The earth was formed 4.5 billion years ago in the privileged cosmic time-series." The same holds for the statement, "The universe began to exist 15 billion years ago with the big bang."

It is important to note that the privileged cosmic time-series comprised by the successive planes of homogeneity does not contain absolute temporal relations. The temporal relations are relative, but privileged. If two events are simultaneous in the privileged cosmic time-series, they are simultaneous *relative to a reference frame whose plane of simultaneity coincides with the plane of homogeneity*.

In this section and the previous sections I have described the temporal notions based on General Relativity that apply to our universe. In the following sections I shall explore the other possible times allowed by General Relativity.

PARALLEL TIME-SERIES

Although the theory of a privileged cosmic time is the sort of physical time that most likely applies to reality, there are other general relativistic theories of time and some of these are taken by physicists more seriously than others. In this section and the following sections, I describe several theories of time, beginning with the ones that physicists think are the most serious candidates for theories that may be found to be applicable to physical reality and ending with the ones they believe are highly unlikely to be confirmed by any observations of physical reality. Although only the theory of privileged cosmic time is widely accepted today as having an application to reality, some physicists believe there may be reason to think that some other temporal notions may also apply to reality.

The concept of parallel time-series is employed by many contemporary cosmologists, but perhaps the most viable employment is in Stephen Hawking's quantum cosmology. Hawking formulates an equation that predicts the unconditional probability with which universes will begin to exist uncaused. This equation is the so-called "wave function of the universe." This equation predicts that many universes will begin to exist uncaused, expand to a maximum radius, and then recollapse and go out of existence. Each of these universes is spatially disconnected from each other universe and each has its own privileged cosmic time-series. If you were inside one of these universes and you traversed every single point in space in that universe, you would find no pathway between that universe and another universe; the spatial points in one universe are spatially related only to other points in that universe and not to any of the points in another universe. Analogously, the times in one series are not temporally related to any of the times in the privileged cosmic time-series of another universe. For example, a certain moment in our universe is not earlier or later than or simultaneous with any moment in the privileged cosmic time-series of any of the other universes. This is implied by their spatial disconnection; if there is no spatial connection, then it is impossible for light signals to be sent from one universe to the next to establish the requisite temporal relations. Furthermore, it is impossible to interlock the planes of simultaneity within one universe with the planes of simultaneity of another universe (to form a common cosmic time) since an interlocking of planes of simultaneity requires a spatial connection. Thus, we have parallel time-series.

The term "parallel" in describing these time-series is purely metaphorical and refers in effect to the pictorial representation of these time-series as two parallel lines. This is an appropriate pictorial representation, since parallel lines (in Euclidean space) never intersect, which represents the fact that the time-series are entirely disconnected from each other.

Physicists today entertain very seriously the idea that there are other disjoint universes, each with their own time-series, but there is not enough hard evidence to make this suggestion one in which we should place a lot of confidence. Apart from the fact that Hawking's theory hinges on some controversial assumptions about quantum gravity, there is an argument (technically called the "argument from decoherence") that the other universes predicted by his equation should not be accorded reality.[1]

TREE TIME

In the 1970s and early 1980s several cosmologists, such as Edward Tryon; Brout, Englert and Gunzig; Atkatz and Pagels; and J. R. Gott developed some cosmological theories that featured tree time. These theories hold that our universe is a branch of a larger, background, empty space–time. The background space–time forms small regions that expand from the background space and form branch-universes. A diagram will show how this works (Figure 7). A and B are origin points of two branch universes. At point A, a small amount of matter forms and explodes in a big bang and a branch-universe then expands from the background space–time and becomes our universe, with all its galaxies and planets. Point B is the origin point of another universe.

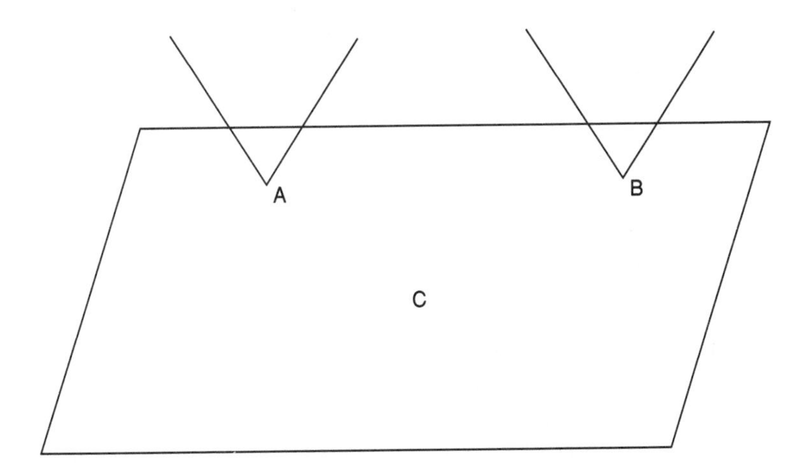

Figure 7 Branching universes
C is the background space–time. A and B are origin points of two universes that emerge from the background space–time.

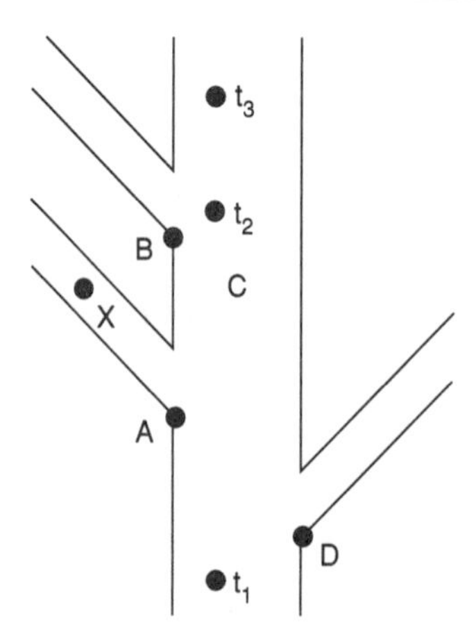

Figure 8 Tree time
C is the time of the background space–time, t_1, t_2 and t_3 are times in the background space–time. A, B and D are origin points of time-series that branch off from C. X is a time later than A.

In this cosmological scenario, there is tree time. Figure 8 represents this time in terms of a picture of a tree trunk with its branches. The trunk-time, C in Figure 8, is the time of the background space–time. The times t_1, t_2 and t_3 are successive times in the trunk-time. The time of our universe, which originates from point A, is the branch that separates off from the main trunk-time at point A and the time of another universe is the branch that separates off from the trunk-time at the later point B. The point X is a time in our branch universe that is later than the time A. The times in our branch universe are later than the times in the trunk-time that occur prior to the origination point A. This is clear from the fact that the events occurring at A are causal effects of some of the events in the background space–time that occur prior to the point A. However, *the time X is not temporally related to the times t_2 and t_3 in the trunk-time* C. The times t_2 and t_3 are later than the time at A, which is a time that belongs both to the trunk-time C and the branch-time that begins from A. But this fact does not imply that the time X has a temporal relation to t_2 and t_3. Time X is temporally unrelated to t_2 and t_3 since X meets none of the conditions required in Einstein's theories of relativity for being temporally related to t_2 and t_3. For one thing, the events occurring at X are neither causes nor effects of the events at t_2 and t_3 (and thus fail to meet one condition of being earlier or later than them). Moreover, the events at time X are not on any planes of simultaneity with the events at t_2 and t_3. In addition to this, the events at X do not belong to any planes of simultaneity that are earlier or later than the planes of simultaneity to which the events at t_2 and t_3 belong.

Thus, we have this interesting situation. The time A is earlier than the time X and earlier than the time t_3, but X and t_3 are temporally unrelated to each other.

Furthermore, the time X has no temporal relation to the times in the other branch universes that originate from the points B and D (see Figure 8). The time A is later than the time D and earlier than the time B, but the times later than A have no temporal relation to the times later than B or D. Such are the peculiarities of tree time.

The reason why there is not adequate evidence that tree time *actually exists* (as opposed to being a mere possibility) is discussed in the next section.

INTERSECTING TIME

Figure 9 represents a cosmological scenario similar to Figure 7, except that the expanding universes intersect. The branch-times that separate off from the trunk-time eventually intersect one

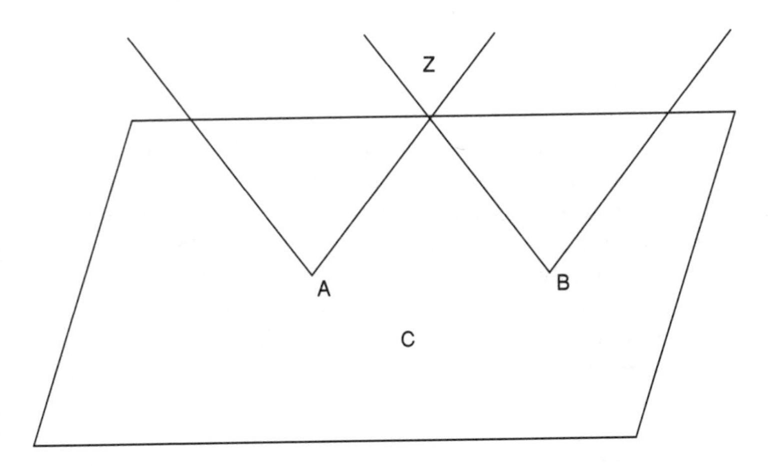

Figure 9 Intersecting universes
A and B are the origin points of two universes from the background space-time C. These two universes intersect, and Z is the region common to both the universes.

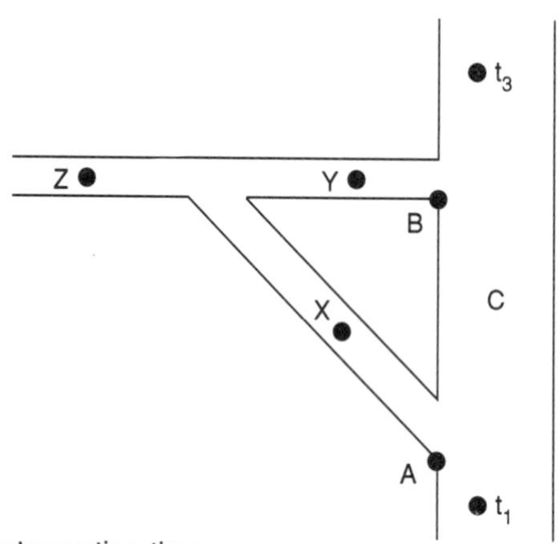

Figure 10 Intersecting time
C is the trunk time. X, Y and Z are times in the branch-times. A and B are the origin points of the two branch-times.

another. In Figure 10 this is represented. The time A that begins one branch-time is later than the time t_1 in the trunk-time and earlier than the time B that begins another branch-time. Likewise, a time Z in the time-series that is formed from the intersection of the A and B branches is later than the time Y and later than the time X, even though X and Y have no temporal relations to each other (see Figure 10). The time Z is also later than the time t_1 in the trunk-time. But the time Z bears no temporal relation to the time t_3 that belongs to the trunk-time.

The question of whether there is empirical reason to believe in tree time or intersecting time depends on whether or not we possess observational evidence that our universe has intersected some other universe. Various technical considerations show that if there are branching universes, as depicted in Figure 7, then they will eventually intersect, as depicted in Figure 9. However, there is no evidence that there is any such intersection and it is likely we would have observed such intersection if it existed. Thus, it seems unlikely that there actually exists either tree time or intersecting time.

SPLITTING TIME

Splitting time is consistent with the General Theory of Relativity, but it is implied by Hugh Everett's "Many Worlds" interpretation of quantum mechanics. Everett's theory implies that the universe is continually splitting into multiple copies of itself, such that each split involves several new time-series that stem from the old time-series. Everett's theory can be applied to a universe that begins with the big bang, with time beginning to split at the instant of the big bang (see Figure 11). This type of time is rarely if ever distinguished from tree time, but they are distinct. Tree time has the property of having a single temporal series, the trunk-time, being the origin point of each new temporal series. But splitting time has the property that new temporal series do not always originate from the same temporal series. As we see from Figure 11, the big bang at t_0 that begins time is such that time immediately splits into two series, A and B. The series A in turn eventually splits into C and D, which in turn branch and so on.

This reveals a further difference from tree time. In tree time, it is possible for the trunk-time to have an infinite future and infinite past. And it is possible for each of the tree-branch times to have an infinite future. But with splitting time, no time-series has an infinite past or an infinite future. Each time-series extends for a finite time before it ends and splits into two (or perhaps more) new time-series.

Each event in each time-series is temporally related to the big

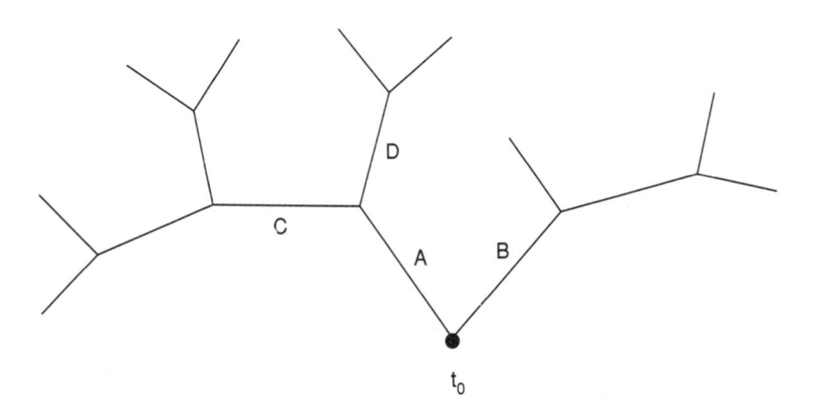

Figure 11 Branching time
The big bang is the time t_0. Time immediately splits into A and B; the time-series A splits into C and D, and so on.

bang. Consider any event in the time-series D. It is clear from Figure 11 that any such event is later than the big bang, since there is a continuous causal sequence extending from the big bang through the series A and through the series C. However, the events in the temporal series D bear no temporal relation to the events in most of the other series; for example, the events in D are not temporally related to the events in the series B and C.

During the 1970s, some physicists believed that Everett's Many Worlds theory was true and that there actually existed splitting time. They believed that Everett's theory of quantum mechanics was the only theory that applied to the universe as a whole and that an infinite number of splitting time-series came into existence at times later than the big bang. However, since the 1980s a more plausible version of quantum mechanics, the so-called "Consistent Histories" theory has been developed which both applies to the universe as a whole and does not imply that time splits. Thus, there is no good reason at present to believe there actually exists splitting time.

CLOSED TIME

Suppose that the universe expands from the big bang, eventually stars and planets form, and then, after the universe reaches a point of maximal expansion, it begins to contract and eventually goes out of existence at the big crunch. According to present observational evidence, there is about a 50 per cent chance that this scenario is true and a 50 per cent chance that the universe will keep expanding forever.

However, the hypothesis that time is closed differs from both of these scenarios, even though the observational evidence that would support the closed time hypothesis is similar to the expansion–contraction scenario. The crucial difference concerns the identity of the big bang singularity (the instantaneous point at which the universe begins to exist) and the big crunch singularity (the instantaneous point at which the universe ceases to exist). If time is closed, then the big bang singularity is identical with the big crunch singularity. The standard model of the universe, based on Friedman's equations, implies that the big crunch singularity is later than the big bang singularity; the big crunch singularity occurs at the last instant of time and the big bang singularity occurs at the first instant of time. But if

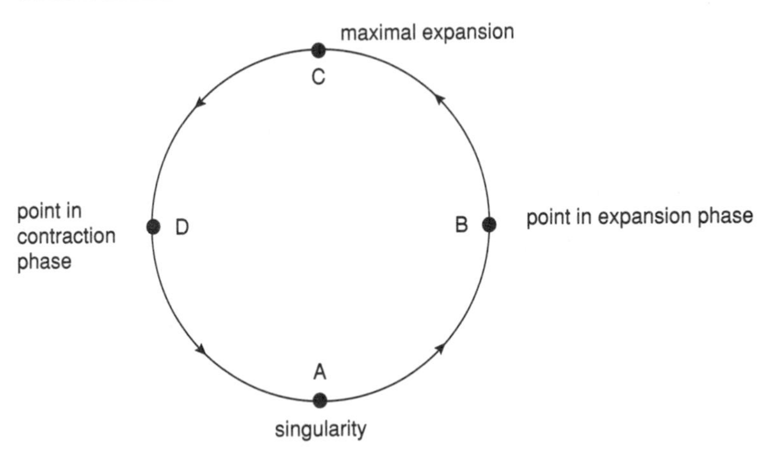

Figure 12 Closed time
The universe is maximally small at A, expands through B and C and then contracts through D and returns to the point A.

these times are identical, then when the universe contracts to the big crunch singularity, the universe does not pass out of existence but instead explodes in a big bang and expands. In this latter case, time neither begins nor ends. Nonetheless, time is finite; there are only a finite number of intervals of time of a given length. Suppose the universe expands for 25 billion years and then contracts for 25 billion years; time contains only 50 billion years, and yet there is no earliest year and no latest year of time.

This will be the case if time is closed, like a circle, rather than open, like a line. Consider Figure 12. A is the singularity. "The big crunch singularity" and "the big bang singularity" both refer to A. B occurs later than A; it is a point during which the universe is expanding. C is the time of maximal expansion and D is a point during the contraction phase.

Is D later than B? In one respect, yes, since the contraction phase is later than the expansion phase. But in another respect, D is earlier than B; D leads to the singularity A, which then explodes and is followed by the expansion phase that contains the point B. So D is both later than B and earlier than B, which seems paradoxical. Furthermore, it appears that D occurs earlier than A, which is earlier than B, which in turn is earlier than C, which in its turn is earlier than D. This implies that D is earlier than itself. But "D is earlier than itself" implies "D occurs earlier than it occurs," which seems to be an implicit logical contradiction.

Some authors, such as W. H. Newton-Smith, attempt to ameliorate this difficulty by distinguishing between two sorts of relations between *instants* of time, *being locally earlier than* and *being globally earlier than*. Newton-Smith says that if an instantaneous event E_1 is locally before an instantaneous event E_2, E_2 will not be locally before E_1. E_1 might be locally before E_2, E_2 locally before E_3, and E_3 might be locally before E_1. This would imply that E_1 is globally before itself. The ordinary notion of beforeness, Newton-Smith says, implies only that an event or time cannot be locally before itself. Since closed time implies only that times or events are globally before themselves, closed time is consistent with our ordinary temporal notions and the appearance of paradox is dispelled.

But this distinction will not work in a general relativistic universe where we take our times as instants or our events as instantaneous events. Newton-Smith is suggesting that t_1 is locally before t_2 if t_2 immediately succeeds t_1, and that t_1 is globally before t_2 if there are other times after t_1 and before t_2. However, if time is continuous, as it is represented to be in the General Theory of Relativity, then no instant of time or instantaneous event is locally before any other instant. Between any two instants t_1 and t_2, there is an infinite number of other instants. Thus, for any two instants t_1 and t_2, if t_1 is before t_2, t_1 is globally before t_2. The distinction works in Newton-Smith's model because he adopts "the fictional assumption that we are dealing with worlds in which there are only a finite number of instants or moments, say four."[2]

However, if we introduce *intervals* of time, then the distinction will work in a general relativistic universe. No interval will be locally before itself, but each interval will be globally before itself. This is the case even if we divide all of time into two equal lengthened intervals. In Figure 12, we may take an interval that extends from point A to point C. This interval is locally before the interval that extends from C to A. Since the later interval is locally before the interval that extends from A to C, it follows that the interval that extends from A to C is globally before itself.

There is no evidence that time is closed. Indeed, the most well-confirmed physical theories imply that time is not closed. In terms of the general relativistic theory of the expansion and contraction of the universe, the theory that is best confirmed is that the big crunch singularity occurs later than the big bang singularity and is not identical to the big bang singularity.

A CYCLICAL HISTORY IN LINEAR TIME

This is Nietzsche's eternal return. This is where events A, B, C, and D keep repeating themselves (see Figure 13), and requires a commitment to substantival time. Suppose A, B, C and D are all the events in a cycle of the universe where it expands and then contracts. Suppose the universe, after it contracts, begins expanding again and repeats exactly its history in the previous cycle of expansion and contraction. Event A in one cycle is identical with the event A in a later cycle; the event possesses all and only the same non-temporal properties and relations in the earlier cycle as it possesses in the later cycle. If each cycle in the expansion–contraction phase were composed of the numerically identical events, but the moments of time were different, then there would be an endless repetition of the identical events. Event A in one cycle would differ from event A in a later cycle only in that A would possess the temporal relation, *occupying moment* M_1 in one cycle and would instead possess the temporal relation *occupying moment* M_5 in the later cycle. This indicates the substantival theory of time would be true.

Why does an endless repetition of the identical events require the substantival theory of time? If the relational theory of time were true, and there were no times distinct from the sets of the events, then there would be no distinction between the occurrence of event A in one cycle and the occurrence of the event A in another (alleged) cycle. If A occurs and then B, C, D and then

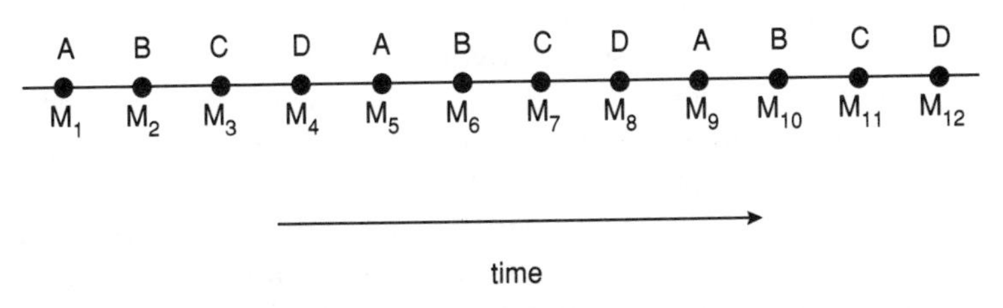

Figure 13 Cyclical history in linear time
A, B, C and D are events that recur endlessly. M_1, M_2 . . . M_{12} are successive moments of time occupied by these events.

A "again", we would have closed time. But in the case of closed time we could not accurately say that "A occurred again," but instead should say that A is globally later than itself in a closed time. In a closed time, the event A occurs only once. Time would be open and linear *only if there is something that distinguished one occurrence of A from another occurrence of A*. If the later occurrence of A occupied a moment M_5 and the earlier occurrence a moment M_1, then there would be a distinction, a difference in the moments that A occupied. Since M_5 is later than M_1, we could say that A occurs twice, once at M_1 and later at M_5.

It might be thought that there could be an endless repetition of the identical events if both the relational theory of time and the tensed theory of time were true. If the tensed theory of time is true, events possess properties of futurity, presentness and pastness. First A possesses presentness, and then B, C and D, and then A once again possesses presentness. It might be said that this provides a means of distinguishing the two occurrences of A. Call the first possession of presentness by A the temporal occurrence T_1 and call the later possession of presentness by A the temporal occurrence T_5. The two occurrences of A are different inasmuch as T_1 is different than T_5. Thus, a cyclical history in linear time is arguably possible if the relational theory of time and the tensed theory of time are both true.

However, this argument (which I earlier championed in "A New Typology of Temporal and Atemporal Permanence"[3] seems to me now to contain a serious logical difficulty. If the event A is present, then A is also past and future. This implies the explicit logical contradiction that if A is present, then A is not present. In order to avoid this contradiction, we need to introduce substantival time in the form of moments. If A-as-occupying-M_5 is present, that is consistent with A-as-occupying-M_1 being past and A-as-occupying-M_9 being future. Thus, cyclical history in linear time requires the substantival theory of time.

There is no evidence that there is a cyclical history in linear time. Indeed, there is positive evidence against it. For one thing, there is no plausible evidence that there are many cycles of expansion and contraction.[4] Furthermore, there is compelling reason to think that even if there were different cycles of expansion and contraction, the different cycles would contain different kinds of events or states. For example, later cycles would contain more radiation than earlier cycles and would expand to greater radii. Further, the second law of thermodynamics ("Disorder always increases") implies that each cycle would be more disordered than the cycle that preceded it. Thus, the concept of a cyclical history in linear time remains a mere conceptual possibility.

A REVERSING HISTORY IN LINEAR TIME

In 1986 Stephen Hawking published an article, "Arrow of Time in Cosmology," in which he argued that the contraction phase of the universe would be the exact reverse of the expansion phase, like a movie running forward (the expansion phase) and then being run backwards (the contraction phase).[5] He and other physicists later realized this theory was false, but the theory is interesting nonetheless since it raises the question about what sort of temporal structure it implies.

In Hawking's 1986 universe, does time reverse, going first forward and then backwards? Or is it merely the case that the events in it occur in reverse order, first forward and then backwards, with time always going forward? Hawking says of this universe that "The contracting phase would be like the time reverse of the expanding phase. People in the contracting phase would live their lives backwards: they would die before they were born and get younger as the universe contracted."[6]

The best way to answer these questions is to select some instantaneous events that occur in this universe. Since time is continuous, there would be an infinite number of instantaneous events, and between any two events there would be an infinite number of other events. Let us consider three of these events, which we shall call 1, 2 and 3, that occupy different positions in the history of the universe. Consider the instantaneous event 1 that represents the big bang singularity, the event 2 that belongs in the expansion phase, and the event 3 that represents the point of the maximal expansion of the universe. Hawking's theory implies that event 1 occurs, event 2 occurs later, event 3 later still, and then events 2 and 1 reoccur but in reverse order. The events occur in this order: 1, 2, 3, 2, 1.

If the substantival theory of time is true, then the second occurrences of 2 and 1 occupy different and later moments than the first occurrences. So *time* does not reverse, but the order of the occurrence of events does reverse. The succession of events is represented as 1, 2, 3, 2, 1, but the succession of temporal moments is represented as M_1, M_2, M_3, M_4, M_5.

But if the relational theory of time is true, can we say that time itself reverses? Suppose we describe the situation as follows. The time t_1 is the unit set consisting only of the instantaneous event 1 and t_2 is the set consisting of the event 2. Can we conclude that time has the order t_1, t_2, t_3, t_2, t_1? On this scenario, time t_2 has the relational property of occurring after t_1 and before t_3 and also has the relational property of occurring after t_3 and before t_1. Suppose we are at time t_2. How do we know where we are in the temporal series? If the next time is t_1, then we know we are at the second occurrence of t_2, but if the next time is t_3, then we know we are at the first occurrence of t_2.

Although this scenario may seem coherent on the surface, it is in fact implicitly self-contradictory. What is meant by "the second occurrence" of t_2? We are implicitly meaning that t_2 occurs *at a later time* than the first occurrence of t_2. But this cannot be the case if t_2 is the identical time in both cases. We are implicitly treating t_2 as an event in time, and we are tacitly introducing substantival time, moments of time distinct from the events, that continues to flow forward, with the moment that contains what we called "the second occurrence of t_2" being later than the moment that contains what we called "the first occurrence of t_2." Thus, it is false that it is coherent to suppose that time can reverse. All that can reverse is the order of the occurrence of the events in substantival time.

If we attempt to combine the relational theory of time with the tensed theory of time, and argue that in this case time can reverse, we would also meet with a contradiction. If t_2 in the contraction phase is present, then it is also true at this time that t_2 is not present, since t_2 in the expansion phase is past. *t_2 is present and not present* is an explicit logical contradiction.

Thus, a reversing history in linear time requires the substantival theory of time if it is to be possible at all. But the present evidence is that it is a mere conceptual possibility. History does not reverse itself.

TIME-TRAVEL INTO THE PAST

The last concept of time I shall discuss is a conception that allows time-travel into the past and travel from the past back again to the present or future. Time-travel was not taken very seriously by most thinkers until 1949, when Kurt Gödel showed that Einstein's General Theory of Relativity allowed for time-travel. Gödel indicated that if the entire material content of the universe is in a state of uniform rotation, then one can take off from the earth in a rocket and travel into the past. Gödel's universe rotates and does not expand, whereas our universe does not rotate and is expanding, and so time-travel of the sort envisaged by Gödel is not possible in our universe. Nonetheless, the fact that time-travel would be possible under certain conditions makes it worth discussing.

The time-travel allowed by Einstein's General Theory of Relativity is most easily seen in terms of a diagram (Figure 14). We will suppose that this universe is exactly like our universe, with the difference that the average smoothed-out mass content in the Gödel universe is rotating and the universe is not expanding. The average smoothed-out mass content is in rotation relative to the paths of freely moving light-rays or material particles. These are the paths of light-rays or particles that are subject to no forces as viewed from a reference frame fixed in the rotating mass content.

The central vertical line in Figure 14 is the world-line of the earth; it is the history of the earth in space and time. The years marked on this line are earth-times, 1990 through 1994. The earth's world-line runs up the central cylinder in Figure 14. Every world-line in this central cylinder that is parallel to the earth's world-line marks off the same time as does the earth's world-line.

A rocket takes off from the earth in 1992 and heads toward the outside of the central cylinder of the universe. During this time, the rocket is going forward in time in the normal way; it goes from earlier to later times, as do other objects in the central cylinder. But once the rocket escapes the central cylinder it begins its time-travel to the past. The reason for this is that outside of this central cylinder, the rotational effects of the universe become predominant and light cones are tipped over, relative to the vertically-oriented light cones inside the central cylinder. In Figure 14, A is a vertical light cone inside the central cylinder and B a tipped-over light cone, outside the central cylinder. Once the rocket escapes the central cylinder, it enters the region where the light cones are tipped over and thus its own forward journey in time becomes "tipped over" or past-directed relative to the earth-

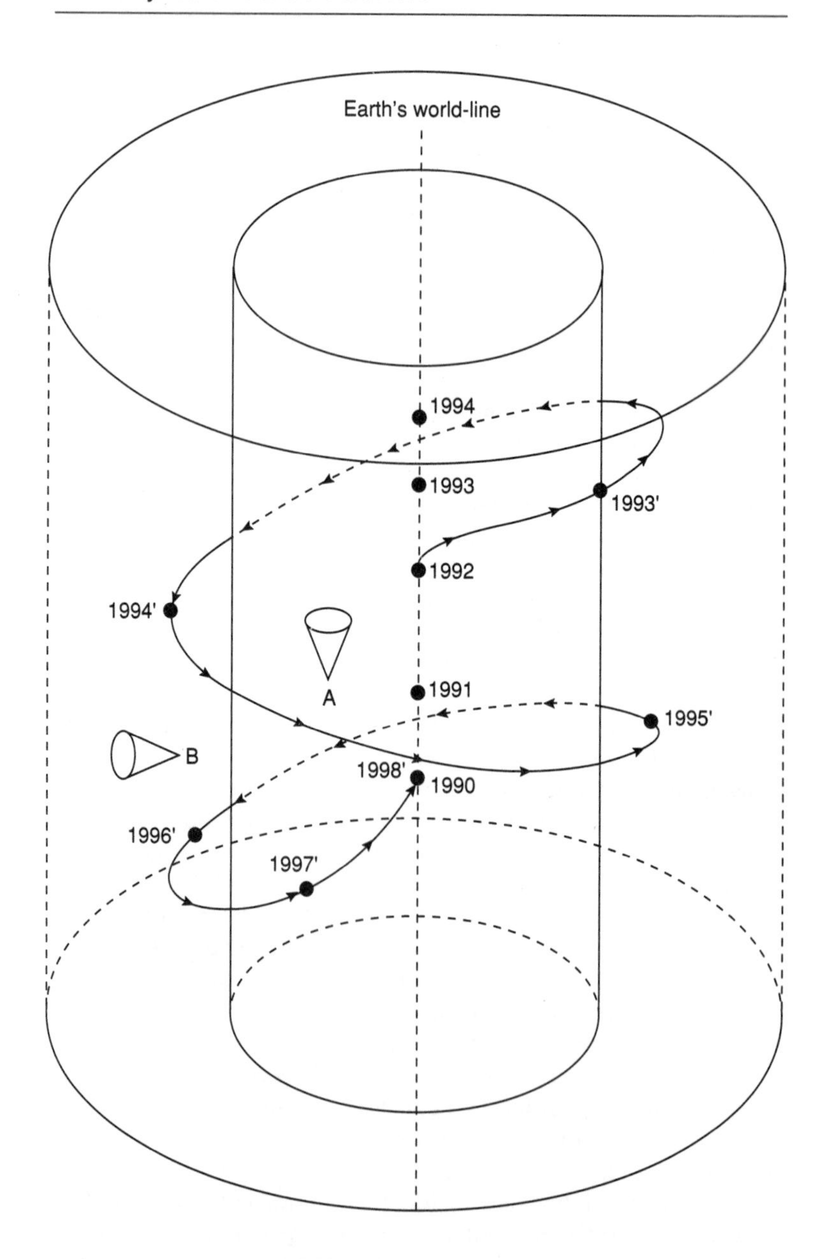

Figure 14 Gödel's universe
(1T) Earth-times: 1990, 1991, 1992, 1993
(2T) Rocket-times: 1993', 1994', 1995', 1996', 1997', 1998'
A and B are light cones.

time. (Gödel's theory implies a gradual transition from the earth's world-line to the region outside the cylinder, but I have made the boundary sharp for ease of exposition and diagramming.)

Eventually the rocket re-enters the central cylinder and lands on the earth, but at some time ago, in the year 1990. So the person gets out of the rocket and finds herself at an earlier point in time than when she left.

Jane leaves the earth in 1992 in a rocket and arrives back on the earth in 1990; this implies she first occupied 1992 (before she took off in the rocket) and later occupied 1990, after she landed. This might appear at first glance to be a contradiction. If Jane is first in 1992 and then *later* in 1990, that might appear to be inconsistent with the fact that 1990 is *earlier than* than 1992.

But Einstein's theory of the relativity of time enables us to make sense of this idea. The time-traveler's experiences of the events in her life must occupy a series of events connected by earlier–later relations that do not always correspond to the earlier–later connections of ordinary earth-time. In earth-time, 1990 is earlier than 1992, but the time-series of the time-traveler is such that she is located (in her time) in 1990 later than she is located in 1992. In the case of our diagram, the time measured by the time-traveler's clock in her rocket is a different time than earth-time. The events on the earth that took place in the year we label "1990" are earlier than 1992 in the earth's time, but these events occur later than 1992 in the rocketeer's time. According to her clock, the events that occur in 1990 in earth-time occur in 1998′, six years after she took off from the earth in her rocket.

Another objection that is sometimes made to the thesis that time-travel is possible is that it requires permanently true statements to become false, which would be a contradiction. If a tenseless statement is true at some time, it is true ever after. For example, "The Battle of Waterloo occurs one year earlier than 1815" is true in 1815 and is true at every later time. Now the objection against time-travel goes like this: "Jane does not land on the earth in a rocket in 1990" is true in 1991 and is true ever after. Thus, this statement cannot become false at a later time and thus it cannot become true at a later time that "Jane lands on the earth in a rocket in 1990." Therefore, Jane cannot time-travel to 1990 since this would make a permanently true statement change from being true to being false.

The response to this objection is that the theory of time-travel can accommodate the permanence of truth. If Jane time-travels to 1990 (in earth-time), then it is changelessly true that "Jane lands on the earth in a rocket in 1990." It is never true that "Jane does not land on the earth in a rocket in 1990." The people living on the earth in 1990 would see Jane land and would listen to her incredulous account of how she time-traveled to 1990 from 1992. They would record this event and we would be able to read about it now, in 1992. Thus, this objection against the possibility of time-travel does not succeed.

But how could it be true in 1991, before she took her rocket journey, that she has already landed back on the earth in 1990? The people in 1991 remember Jane landing in 1990, so they know that it is true that "Jane landed on the earth in 1990." If they are aware that she time-traveled, they are also aware that "Jane will leave the earth in 1992 on her journey back to 1990." They know that Jane has not yet left, but has already landed. This may sound paradoxical, but it is not paradoxical if we take into account the fact that the time-series of Jane is different from the time-series of people on the earth. In Jane's time-series, she lands *after* she takes off. Since the rocket's journey takes place in Jane's time-series, the fact that it lands after it takes off in her time-series is sufficient to make the journey possible. In the earth's time-series, the rocket lands *before* it takes off, but the rocket's complete journey does not take place in the earth's time-series and hence "the rule that the beginning of the rocket's journey is before its termination" does not apply within the framework of earth-time.

Another objection to time-travel is the famous "killing your parents" objection. If time-travel is possible, then I can travel into the past and kill my parents before they conceived me. Suppose the point marked 1990 on the central line in Figure 14 is not 1990 but the year 1920. I get out of my rocket in the year 1920 and kill my parents. But if I killed my parents, I would never have been born. It would be false that I exist. But if I kill my parents, I must exist. So we would have the contradiction "I exist and do not exist."

It is interesting to note that a problem of this sort worried Kurt Gödel when he developed his theory of time-travel. He tried to solve it by saying that the rockets required to return to the past would need more fuel than we presently know how to put into rockets.

But the "not enough fuel" ploy is not a solution to a logical problem. It is logically possible that we develop the right kind of rockets and fuel and so we would still be faced with the same logical contradiction, namely, that it is logically possible that we could do something logically impossible.

The real answer to the "killing one's parents" objection is that there are logical constraints on everything, including time-travel. I cannot do what is logically impossible when I am not time-traveling, and similarly, I cannot do what is logically impossible when I am time-traveling. It is impossible for me to do something that would contradict the fact that I am alive and time-traveling. Thus, I can travel into the past and attempt to kill my parents, but I would not succeed in this. I may fire bullets at them, but the bullets would miss.

This point can be stated in an even stronger way. The reason that you cannot travel into the past and kill your parents before you were born has nothing to do with time-travel but is a matter of logic. Even if time travel is impossible, it is still impossible to kill somebody's parents before he is born. If Alex is now 10 years old, it is impossible that somebody who lived twenty years ago killed Alex's parents then.

It is worthwhile further developing the point I made earlier, that time-travel requires more than one time-series. If there is a second time-series, then if you are located at any given time on a certain world-line, say the world-line in our diagram that represents the history of the earth from 1990 to 1994, then you can go "elsewhere" in time than to the immediately later time in earth-time. You do not do this by reversing direction in earth-time but by entering a second time-series that eventually overlaps the original time-series (earth-time). Now let us represent the first time-series, the earth-time in which we now exist, as 1T and the second time dimension, in which Jane's travels take place, as 2T. Suppose Jane takes off in a rocket in 1992. Look again at Figure 14. As long as you remain within the inner cylinder, you belong to the same time-series as does the earth. ("Earth-time" is in a sense a misnomer since this time-series extends far out into space, including the whole solar system, the Milky Way galaxy and many surrounding galaxies.) As long as your rocket remains within the inner cylinder, you remain within this time-series, 1T as we may call it. But as you pass outside of the cylinder, you enter a second time series, 2T. Suppose it takes a year to reach the edge of the cylinder, so that it is 1993 in 1T as Jane reaches the edge. As Jane passes outside the cylinder, the clock on board her rocket reads 1993′, then 1994′, then 1995′ and finally 1998′, which is the time at which she lands again on the earth and finds that the earth-clocks read 1990.

The interesting correspondence is between 1998′ in the rocket's time dimension and the year 1990 in earth-time. What is the relationship between these two times? Suppose Jane steps out of the rocket and finds her 1990 self lying in a hospital bed with a broken leg. Suppose that the rocketeer's time 1998′ is simultaneous with 1990 in earth-time. If two times are simultaneous, they are the same time. This is built into the concept of simultaneity, much as "two events are simultaneous" means the two events occur at the same time. But if the two times are simultaneous, this implies that at one and the same time, Jane's leg is broken and is not broken, which is a contradiction. The rocketeer's time 1998′, then, cannot be the same time as the earth-time 1990. They cannot be simultaneous, since that would mean they are the same time. There must be some other relation between them. There is no *temporal* relation, given that all temporal relations are either relations of simultaneity, being earlier than or being later than. Rather, there is a relation of *dual occupancy*. The events on the earth that the rocketeer interacts with in her time 1998′ *occupy her time, 1998′* and these same events also *occupy the earth-time 1990*. The events occupy times in both of the time-series, and thus the two times are "dually occupied" by these events. For example, perhaps the 1988′ Jane disguises herself as a doctor and talks with her 1990 self that is laid up in bed. This conversational interchange occupies both the 1998′ time and the 1990 time.

These ideas are based on the familiar idea in Einstein's Special Theory of Relativity that each world-line has its own proper time. The proper time of a world-line is just the time measured by an actual or hypothetical clock (e.g., light clock) that travels along that world-line. The rocketeer's proper time is measured by her clock, which ticks off the years 1997′, 1998′, etc., and the earth's proper time is measured by its clock, which ticks off the years 1990, 1991, 1992, etc. The "dual occupancy" relation is based on the idea that the same event can belong to two different proper times, which occurs whenever two world-lines cross each other (regardless of whether or not this involves time-traveling).

Considerations such as these show how the tensed theory of

time is compatible with time-travel. It is often objected that time-travel requires the tenseless theory of time, which implies that successive events are equally real, that they have the same ontological status. According to the tensed theory of time, past events no longer exist and future events do not yet exist. Thus, it would seem impossible for a person at present to travel into the past. For then the person would be traveling to a time that no longer exists. Certainly, it is impossible for a rocket to land at a place that does not exist any longer!

However, this objection can be answered in terms of the tensed interpretation of Einstein's Special Theory of Relativity, which I outlined in the previous essay. This interpretation relativizes reality to a reference frame. In the case at hand, what is real relative to the reference frame of the rocketeer need not coincide with what is real relative to the reference frame of the earth. Relative to the reference frame of the earth in 1992, the events on the earth in the year 1990 are no longer real. But relative to the reference frame of the rocket in the rocket's year 1998′ (when the rocket has landed on the earth), these same events are presently real. The events that occur on the earth during the time the rocketeer has landed are present relative to the rocketeer's time 1998′ but are past relative to the earth-time of 1992.

One can also time-travel from the past year 1990 back to the present year 1992 or even to a later year. In the Gödel universe, the rate at which one travels through time depends on how fast the rocket travels. The closer to the speed of light the rocket travels, the faster you travel through time. If at a certain speed, it takes one year in the rocketeer's time to travel from the earth-time 1990 to the earth-time 1994, then at a greater speed it would take only half a year. Thus, the rocketeer who lands on the earth at the earth-time 1990 can return to 1994 in one year of rocket-time if she flies sufficiently close to the speed of light.

CONCLUSION

We have explored various theories of physical time. However, we have seen that none of these theories applies to our universe except for the theory of privileged cosmic time. It will be recalled that this time consists of planes of simultaneity that coincide with the planes of homogeneity of our universe. From the reference frame of these planes, the universe appears to expand uniformly, to contain evenly distributed matter and to look the same in every direction (on a very large scale). This reference frame gives us the truth about physical time and change in our universe. It tells us that it is 15 billion years since the universe evolved without cause from literally nothing at all and that we are now expanding into an unknown future.

NOTES

1 For a further discussion of Hawking's theory, see my "The Wave Function of a Godless Universe," in W. L. Craig and Q. Smith (1993) *Theism, Atheism and Big Bang Cosmology*, Oxford: Clarendon Press.
2 W. H. Newton-Smith (1980) *The Structure of Time*, London: Routledge and Kegan Paul, p. 58.
3 Q. Smith (1989) "A New Typology of Temporal and Atemporal Permanence," *Noûs* 23.
4 See W. L. Craig and Q. Smith (1993) *Theism, Atheism and Big Bang Cosmology*, Oxford: Clarendon Press, chapter 4 and appendix A.
5 S. Hawking (1986) "Arrow of Time in Cosmology," *Physical Review* D 32: 2489–95.
6 S. Hawking (1988) *A Brief History of Time*, New York: Bantam Books, p. 150.

GLOSSARY OF TERMS

Absolute simultaneity Two events E_1 and E_2 are absolutely simultaneous if a two-termed relation of simultaneity obtains between them. This relation is expressed as "E_1 is simultaneous with E_2."

Absolute space The container of all bodies, in relation to which bodies possess their real lengths, motions and states of rest.

Closed time If time is closed, then time is finite but neither begins nor ends. Time will be represented by a circle, and no point on the circle will be identified either as the beginning or end of time.

Cosmic time The time of the universe as a whole, as it is conceived in the General Theory of Relativity. Each time in a cosmic time-series is a set of interlocking planes of simultaneity that are maximally extended, i.e., extended throughout the universe.

Cyclical history in linear time Time is linear if it is represented by a line, with the time-direction being one direction along the line. There is a cyclical history in linear time if the same series of events keeps repeating itself, so that the events occur as follows: A, B, C, A, B, C, A . . .

The ether An invisible medium that scientists used to think to be at rest in absolute space. The ether was thought to be the medium through which light-rays travel, much as sound travels through the medium of air.

Homogeneity The universe is homogeneous on a large scale if matter is evenly distributed throughout the universe. The universe is homogeneous on the scale of super-clusters of galaxies. A galaxy is a cluster of stars and a super-cluster of galaxies is a cluster of clusters of galaxies.

Intersecting time If two branch-times S_1 and S_2 separate from the trunk-time S and eventually intersect, then this will be an instance of intersecting time.

Isotropy The universe is isotropic if the universe will look the same in every direction and if this is true from any place in the universe. Our universe is isotropic on a large scale, i.e., the scale of super-clusters of galaxies.

Light cone The light cone of an event E consists of a past part, which contains all the events that could causally affect E, and a future part, which contains all the events that E could causally affect. All events outside of E's light cone cannot causally affect E and E cannot causally affect these events.

The light signal method This is a method for determining which events are simultaneous with a given event E. At noon I send out a light signal; it bounces off an event E_1, and returns to me at 2 o'clock. It follows that E_1 is simultaneous with the event E that consists of the dial hands on my watch pointing to 1 o'clock. Thus, E_1 belongs on the plane of simultaneity that runs through the event E.

Parallel time-series There are two parallel time-series if there is one series S_1 that contains times, such that none of these times is earlier than, later than or simultaneous with the times in another time-series S_2. This will be the case if there are two separate universes that are spatially disconnected from each other; each universe will have its own time-series.

Plane of simultaneity All the events simultaneous with some given event E (determinable by the Light Signal Method) form a plane of simultaneity that runs through E.

The Principle of Relativity The laws of physics are the same for all reference frames.

The Principle of the Constancy of the Velocity of Light The velocity of light (186,284 miles per second) is always measured to be the same, regardless of the state of motion or rest of the body from which the measurement is made.

Privileged cosmic time-series If the universe is homogeneous and isotropic, then there is a privileged cosmic time-series. This cosmic time-series consists of planes of simultaneity where the density of matter is constant.

Reference frame A point in space and time relative to which temporal and spatial relations obtain.

Relative simultaneity Two events E_1 and E_2 are relatively simultaneous if a three-termed relation of simultaneity obtains among the two events and a reference frame. This relation is expressed as "E_1 is simultaneous with E_2 relative to the reference frame R."

Splitting time Splitting time differs from tree time. Tree time has a single time-series, the trunk-time, from which the other series branch off. But splitting time has no trunk-time. Rather, there is an initial instant (e.g., the instant of the big bang) that immediately evolves into two time-series S_1 and S_2. S_1 subsequently splits into further time-series and S_2 also splits into further time-series, and so on.

Tree time There is one time-series S that is like the trunk of a tree, and different time-series, S_1, S_2, S_3, S_4, etc., separate off from the trunk-time S. Each of the branch time-series, S_1, S_2, S_3 and S_4 is temporally related to the trunk-time S. But the times in the branch time-series S_1, S_2, S_3 and S_4 are not temporally related to each other.

STUDY QUESTIONS

1. What is the Michelson–Morley experiment?
2. Einstein's argument for the Special Theory of Relativity had the logical form of "(1) If P, then Q. (2) Not-Q. (3) Therefore, not-P." How can this argument be spelled out in terms of statements about absolute space and light?
3. What is Newton's principle of the addition of velocities? How is it inconsistent with the behavior of light?
4. Using light clocks as examples, explain why time goes slower on a body that is moving, relative to some body that is taken to be at rest.

5 How does the example of the train and the lightning flashes show that simultaneity is relative?
6 If you had a flashlight, how would you use it to construct a plane of simultaneity by the Light Signal Method?
7 Explain the argument that the Special Theory of Relativity implies the tenseless theory of time. How can this argument be countered by the defender of the tensed theory of time?
8 State in a simplified form the basic equation of the General Theory of Relativity.
9 Draw a diagram of cosmic time.
10 How does splitting time differ from tree time and intersecting time?
11 Tell a story of time-travel that is consistent with Gödel's universe.

FURTHER READING

Craig, William Lane (1990) "God and Real Time," *Religious Studies* 26: 335–47.
Craig's paper presents a good criticism of Einstein's theory that time is relative and argues that real time is not relative.
*Grünbaum, Adolf (1973) *Philosophical Problems of Space and Time*, Dordrecht: D. Reidel.
The classic work on the philosophical interpretation of the notion of time in Einstein's Special and General Theories of Relativity.
Hawking, Stephen (1988) *A Brief History of Time*, New York: Bantam Books.
An introduction to the theory of time in Einstein's Theories of Relativity and to Hawking's new ideas about the nature of time.
*McCall, Storrs (1994) *A Model of the Universe*, Oxford: Clarendon Press.
This book presents a novel and ingenious integration of the tensed theory of time with Einstein's Special Theory of Relativity.
*Putnam, Hilary (1975) "Time and Geometry," in his *Philosophical Papers*, Cambridge: Cambridge University Press.
Putnam presents the argument that Einstein's Special Theory of Relativity entails the tenseless theory of time.
Salmon, Wesley (1975) *Space, Time and Motion*, Ann Arbor: University of Michigan Press.
Salmon presents a conceptually clear and insightful introduction to Einstein's Special Theory of Relativity in some of the chapters of this book.
Sklar, Lawrence (1974) *Space, Time and SpaceTime*, Berkeley: University of California Press.
This is perhaps the best comprehensive introduction to Einstein's Special and General Theories of Relativity.
*Smith, Quentin (1993) *Language and Time*, New York: Oxford University Press.

In the last chapter, Smith argues that Einstein's Theories of Relativity are not about time but about observable physical changes; Smith argues that real time is an absolute metaphysical time, the study of which is the province of philosophy rather than physics.

Smith, Quentin and Craig, William Lane (1993) *Theism, Atheism and Big Bang Cosmology* Oxford: Clarendon Press.

This book contains interpretations of the notions of time in Einstein's theory of relativity and in Hawking's quantum cosmology.

The following series of articles by V. Alan White and Michael Cohen comprise a clear and illuminating debate about the interpretation of the relativity of simultaneity in Einstein's Special Theory of Relativity, with special attention to the "train and lightning-bolt" thought-experiment.

Cohen, Michael (1989) "Simultaneity and Einstein's *Gedanken-experiment*," *Philosophy* 64: 391–6.

This article challenges some misconceptions about the Special Theory of Relativity, but ends by claiming that Einstein bungled his "lightning-bolt" thought-experiment about the relativity of simultaneity.

White, V. Alan (1991) "Cohen on Einstein's Simultaneity '*Gedanken-experiment*,' " *Philosophy* 66: 245–6.

A response to Cohen's 1989 article in which White argues that Cohen collapses the perceptual and metrical senses of simultaneity.

Cohen, Michael (1992) "Einstein on Simultaneity," *Philosophy* 67: 543–8.

A reply to White's 1991 article in which Cohen argues that Einstein's thought-experiment involving trains and lightning-bolts contains a logical inconsistency.

White, V. Alan (1993) "Relativity and Simultaneity Redux," *Philosophy* 68: 401–4.

A rejoinder to Cohen's 1992 article in which White argues that Cohen uses an unorthodox sense of the relativity of simultaneity to criticize Einstein's "train and lightning-bolt" thought-experiment.

Index

0996